Plants as Factories for Protein Production

Plants as Factories for Protein Production

Edited by

Elizabeth E. Hood

ProdiGene, Inc.,
College Station,
Texas, U.S.A.

and

John A. Howard

ProdiGene, Inc.,
College Station,
Texas, U.S.A.

KLUWER ACADEMIC PUBLISHERS
DORDRECHT / BOSTON / LONDON

A C.I.P. Catalogue record for this book is available from the Library of Congress.

ISBN 1-4020-0843-0

Published by Kluwer Academic Publishers,
P.O. Box 17, 3300 AA Dordrecht, The Netherlands.

Sold and distributed in North, Central and South America
by Kluwer Academic Publishers,
101 Philip Drive, Norwell, MA 02061, U.S.A.

In all other countries, sold and distributed
by Kluwer Academic Publishers,
P.O. Box 322, 3300 AH Dordrecht, The Netherlands.

Printed on acid-free paper

Printed in the Netherlands.

CONTENTS

Part III. Production Issues

INTRODUCTION TO MOLECULAR FARMING

John A. Howard and Elizabeth E. Hood

ProdiGene
101 Gateway Boulevard
College Station, TX 77845, USA

For centuries, people have used plants to extract compounds for pharmaceutical and industrial purposes. In some countries, plants are still a major source of pharmaceuticals and enzymes. However, with the advent of recombinant DNA technology, the situation is changing. The ability to clone your favorite gene and produce your favorite protein in your favorite expression system, gives flexibility never before imagined. This flexibility is needed in order to utilize the exponential number of useful genes identified in genomics programs. It is highly unlikely that a single production system could ever meet all the demands for all products.

Each protein production system has unique characteristics. While no one system offers the ultimate solution, some generalizations can be made. Pharmaceutical proteins are often produced in animal cell culture systems. The need to ensure protein integrity relies on the ability of cell culture systems to process proteins similarly to the native form. Some animal proteins require unique glycosylation that is not performed in non-animal systems. Also, for unknown reasons some animal proteins do not express well in non-animal systems. Therefore, animal cell culture or transgenic animals are the system of choice when these factors are critical. This represents one end of the spectrum where protein integrity is the major factor and cost is secondary. The other end of the spectrum is represented by microbial systems, which are preferred when cost is the major factor. They have the limitation that they do not always work well for animal proteins but are the preferred source for most industrial enzymes.

Some critical factors in determining the system of choice are: cost of goods, initial capital outlay, time to scale up, protein integrity, presence of human pathogens, infrastructure in place, and potential for improvement. Table 1 shows a qualitative assessment of the available production systems with regard to these various factors. These issues are discussed in more detail in the various chapters in this book. This assessment illustrates that plants do offer advantages over other existing systems in several categories. The major drawback is that plants are not used today, and information on plant systems is still in the early stages of development. Nevertheless, though there are exceptions, we believe plants do offer advantages over other existing systems.

The plant kingdom comprises great diversity. Table 2 shows some examples of plant systems that are under evaluation for molecular farming. Several systems including cell culture, forage, grain, fruit, and vegetables are being evaluated. In general, grain crops show the most promise. However, depending on the needs, other plant systems show distinct advantages. While it is not expected that one plant system will meet all requirements for all products, the diversity of plants introduces a new dimension into protein production capabilities. Unlike other types of systems, plants are used in our daily lives for food, fuel, feed, and fiber. These plant-derived compounds are processed to obtain the functional products that we normally use. By using plants as factories, a paradigm shift is occurring (Figure 1), showing that in this new case, the potential exists to make the protein in the plant and then directly process the material into the final product. This puts these new recombinant protein products directly into our daily lives. This could be in the form of orally delivered vaccines or processing enzymes that are engineered into the crop, and then will be used on the crop in later downstream processing steps. Several of these examples are discussed in more detail in the following chapters to illustrate their advantages and utility.

The chapters in this book are arranged into two major sections: systems and product categories, with supporting material for production, legal, and regulatory. We have included chapters representing the production systems currently used or in development. The first commercial products from transgenic plants have recently entered the market. We expect to see many more in the near future. This book will give some insight into what is needed to take a product to the market and the potential utility for the future.

This volume on protein production in transgenic plants provides an up-to-date review of the products, systems and process elements necessary to achieve successful production of proteins from these systems. The chapters are written by the recognized leaders in the field. We thank them for their expert contributions to this timely volume. We also gratefully acknowledge the diligent efforts of Lisa Baker in formatting the camera-ready copy. The editors hope that this book will be useful to those practicing this art as well as those teaching the state of this art to the students who will be the users and recipients of the technology produced by this new industry.

John A. Howard
Elizabeth E. Hood

Table 1. Protein Production Systems

	Bacteria	Yeast	Animal Cell Cultures	Transgenic Animals	Plants
Cost of goods	+++	+++	+	++	+++
Initial capital outlay	++	++	+	+++	+++
Time to scale-up	++	++	+	+++	+++
Protein integrity	+	+	+++	+++	++
Human pathogens	++	++	+	+	+++
Infrastructure in place	+++	+++	+++	+	+
Potential for improvement	+	+	++	+++	+++

\+ Worst of the systems available
++ Better than some, not as good as the best
+++ Best of the systems available

Table 2. *Plant Protein Production Systems*

	Viral System	Cell Cultures	Tobacco Leaves	Forage Crops	Monocot Grains	Dicot Grains	Fruits and Vegetables
Cost of raw material	+++	+	++	+++	+++	+++	+
Purification	++	++	++	+++	+++	+++	+
Scale-up	+++	+	+++	+++	+++	+++	+++
Infrastructure	+	+	+	++	+++	+++	++
Use in food and feed	+	+	+	++	+++	+++	
Ease in laboratory	+++	++	+++	++	++	+	+
Storage and transportation	+	+	+	+	+++	+++	+

Figure 1.

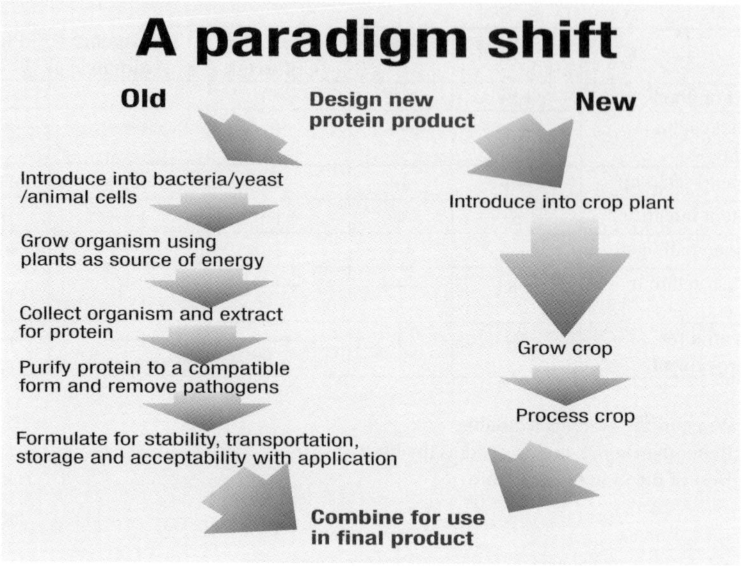

A paradigm shift
Old
Design new protein product
New
Introduce into bacteria/yeast /animal cells
Grow organism using plants as source of energy
Collect organism and extract for protein
Purify protein to a compatible form and remove pathogens
Formulate for stability, transportation, storage and acceptability with application
Introduce into crop plant
Grow crop
Process crop
Combine for use in final product

Part 1
Plant Production Systems

VIRAL VECTOR EXPRESSION OF FOREIGN PROTEINS IN PLANTS

Laurence K. Grill[1], John Lindbo[1], Gregory P. Pogue[1] and Thomas H. Turpen[2]

[1]Large Scale Biology
3333 Vaca Valley Parkway
Vacaville, CA 95688, USA

[2]Eliance Biotechnology, Inc.
P.O. Box 36391
Dallas, TX 75235-1391, USA

INTRODUCTION

It is now fairly routine to engage plant viruses to express foreign proteins in plants. Plant viruses have several features that make them quite useful for vectoring foreign genes into whole plants. The majority of viruses infecting higher plants have single-stranded, positive-sense RNA genomes. Infectious transcripts can be synthesized *in vitro* from full-length cDNA clones to study RNA virus biology, develop methods of disease control and construct plant expression vectors (Goldbach and Hohn, 1997; Scholthof *et al.*, 1996). Tobamoviruses are among the most studied and well-understood viruses and represent superbly efficient genetic systems (Dawson *et al.*, 1986; Dawson *et al.*, 1990; Dawson, 1992). Vectors based on the tobacco mosaic virus (TMV) were among the first to be developed (Donson *et al.*, 1991) and have several advantages for novel applications in the expression of foreign sequences in plants. These advantages include speed, high expression levels, broad host-range, and controlled containment, as transmission occurs only by mechanical means.

ATTRIBUTES OF CURRENT TMV VECTORS

The current TMV vectors carrying foreign genes are able to move rapidly and systemically in plants (generally 7-10 days post-inoculation). The level of expression is high, as the vector is derived from wild-type TMV, which is able to produce more virus-encoded protein per infected cell than any other known plant virus [up to 10% of dry weight in TMV-infected tobacco plants (Coperman *et al.*, 1969)]. Foreign proteins and peptides produced

E.E. Hood and J.A. Howard (eds.), Plants as Factories for Protein Production, 3–16.

systemically in plants by TMV vectors can accumulate to 5% of total soluble protein (Kumagai *et al.*, 1993; Kumagai *et al.*, 1995; Kumagai *et al.*, 2000; Pogue *et al.*, 1997; Turpen *et al.*, 1995). The levels of foreign proteins expressed from viral vectors are generally much higher than that obtainable from transgenic plants or plasmid DNA-based transient expression systems.

With tobamovirus vectors, it is possible to direct the foreign protein to various subcellular locations including the endomembrane system, the cytosol or organelles by utilizing subcellular targeting signal peptides. The subcellular targeting of foreign proteins or peptides can maximize the specific activity and stability, and facilitate the purification of the foreign polypeptides (Pogue *et al.*, 1997; Turpen *et al.*, 1997).

TMV expression vectors can also be used as research tools to study plant biosynthetic pathways, screen gene libraries and express proteins toxic to plant and non-plant systems. Examples include: 1) The carotenoid biosynthetic pathway in plants, which has been altered by up- or down-regulating enzymes responsible for the synthesis of key isoprenoid intermediates (Kumagai *et al.*, 1995). In one case, the levels of intermediate products were altered up to 50-fold by expressing sequences in the plant cytoplasm in either sense or anti-sense orientation. These dramatic alterations of secondary metabolite accumulation can only be achieved by regulated induction in vegetative tissues, and would, most likely, be lethal if constitutively expressed using transgenic technologies. 2) Peptides or proteins such as antimicrobial peptides, animal hormones and growth regulators that would be predicted to be toxic or accumulate poorly in microbial or transgenic expression systems have been expressed in plants from TMV vectors (Turpen *et al.*, 1997).

The greatest advantages in using plant viruses as expression vectors are the characteristics of their hosts and the virus-host interaction. Transfected leaves are one of the most economical sources of biomass for commercial product development and can be inexpensively scaled to meet production requirements. Recombinant TMV vectors have been used in multiple outdoor field trials since 1991. Expression characteristics, host range, persistence in the environment and large-scale plant extraction procedures have all been developed to proceed into commercial production (Grill, 1992). Multi-ton extraction of tobacco tissue grown in the field has resulted in the purification of kilogram (kg) quantities of recombinant viruses for development of vaccines and anti-microbial peptides. Because TMV is mechanically spread in nature (Zaitlin and Israel, 1975), recombinant vectors are contained in the inoculated fields (Grill, 1992). Additional safety in containment is provided by the fact that TMV is not field-transmitted through pollen or seed.

PHYSICAL CHARACTERISTICS OF TMV

TMV, the type member of the tobamovirus group, has a single-stranded, positive-sense, RNA genome of about 6400 nucleotides. The genome contains at least four functional open reading frames (ORFs) and employs two distinct strategies for protein expression; read-through of an amber stop codon and production of subgenomic mRNAs. The 5' ORF of TMV is translated from the genomic RNA resulting in the accumulation of a 126 kDa protein. A translational read-through of an amber stop codon occurs at a frequency of 5-10% resulting in synthesis of a 183 kDa protein instead of the 126 kDa protein. TMV also encodes a 30 kDa protein required for cell-to-cell movement (MP; movement protein) and a 17.5 kDa structural protein (CP; coat protein). MP and CP genes are expressed from separate subgenomic mRNAs [Fig. 1A; (Dawson and Lehto, 1990)].

During the replication cycle, the 126 and 183 kDa proteins are first translated from the genomic RNA. These proteins (and perhaps other host-encoded proteins) comprise the viral RNA-dependent RNA polymerase (RdRP) activity. The viral RdRP synthesizes a full-length "minus-sense" copy of the genome. Minus-sense RNAs are then used by the RdRP as a template for the amplification of plus-sense genomes and mRNAs for MP and CP synthesis (Dawson and Lehto, 1990). The MP, produced early in the infection cycle (2-10 hours post infection, hpi, in tobacco protoplasts), facilitates the movement of progeny viral RNA genomes from initially infected cells to adjacent cells by modifying the size exclusion limit of plasmodesmata junctions (Deom *et al.*, 1992). Later during the infection (6-24 hpi in tobacco protoplasts), production of the genomic RNA, synthesis of CP mRNA and CP become the major synthetic events in the virus replication cycle. CP is produced at very high levels and is required for long-distance, systemic movement of the virus in plants (Dawson, 1992). As much as 70% of cellular translation is devoted to CP production during the peak of viral synthesis (Siegal *et al.*, 1978). CP subunits initiate encapsidation of the virus genomic RNA by binding to the origin of assembly, a specific RNA sequence of about 75 nucleotides that lies within the MP ORF (Dawson and Lehto, 1990). Once formed, virus particles align in paracrystalline arrays in the cytoplasm of infected cells. Because of the stability of TMV virions and the high levels of genomic RNA and CP accumulation, gram quantities of TMV can be easily purified per kg of TMV-infected tobacco tissue (Gooding and Herbert, 1967; Zaitlin and Israel, 1975).

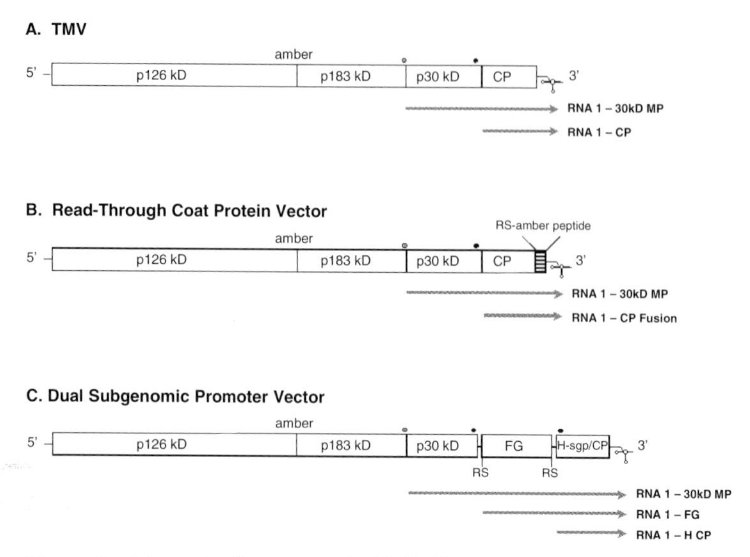

Figure 1. *Schematic diagrams of the TMV U1 genome (A), generalized read-through coat fusion vector (B) and dual subgenomic vector (C). Open reading frames are indicated by open boxes, non-translated regions by lines and 3' tRNA-like structure is schematically drawn as a cloverleaf structure. The amber stop codon and associated contextual CARYYA sequence responsible for translational read-through at the 126 kDa/183 kDa junction is indicated by "amber". Locations of subgenomic promoters are indicated by circles. Subgenomic RNA products are drawn as lines with arrows followed by gene sequence translated from each: 30k MP - 30 kDa movement protein; CP - coat protein; H-sgp/CP - heterologous tobamovirus subgenomic promoter and coat protein ORF; H CP - heterologous tobamovirus coat protein; FG - foreign gene. RS indicates placement of unique restriction endonuclease sites.*

DESIGN OF TMV-BASED VECTORS

The understanding of TMV gene expression at the molecular level has led to the development of two general expression vector designs; CP fusion vectors and dual subgenomic vectors (Fig. 1B,C). These vectors were designed to allow for expression of either peptides or complex proteins. Usually, inserted sequences are expressed throughout the systemic infection of a plant inoculated with synthetic transcripts.

Contrary to initial predictions, foreign sequences propagated in TMV-based vectors accumulate very few point mutations during multiple passages in whole plants (Kearney *et al.*, 1993). However, there can be substantial differences between the longevity of expression of different foreign genes during serial passage. Therefore, care must be taken to monitor the virus population upon passage since each recombinant virus will eventually lose inserted foreign sequences. Mechanisms responsible for the variable instability of foreign sequences are not fully understood. In some cases, foreign RNA sequences are rapidly deleted due to a recombinational event (Turpen *et al.*, 1997).

CP-fusion vectors

One major advantage of the TMV-based vectors is that the molecular structures of the TMV CP and intact virion have been refined at 2.9 Å from X-ray diffraction studies (Namba *et al.*, 1989) providing a model carrier for presentation of foreign sequences. Each virion consists of a single molecule of genomic RNA helically encapsidated by approximately 2100 copies of CP. The core of a single CP subunit contains a bundle of four anti-parallel α-helices projecting away from the RNA binding site. The CP N- and C-termini as well as a loop connecting two α-helices are on the virion exterior surface. Each of these regions may accept a peptide sequence depending on the chemical nature of the individual peptide (Fitchen *et al.*, 1995; Hamamoto *et al.*, 1993; Sugiyama *et al.*, 1995; Turpen *et al.*, 1995). Successful fusions, which do not significantly interfere with virion assembly and disassembly processes or induce a necrotic response in the host plant, may be purified in large quantities as TMV virions. TMV virions can be readily purified in large quantities by relatively simple procedures and equipment. Thus the design of vectors that fuse peptides to the virion can be an effective means of obtaining large quantities of products. If desired, peptide-CP following purification.

To construct a read-through fusion, the nucleotide sequence CARYYA (R=Purine, G or A; Y=Pyrimidine, C or U/T) can be placed immediately after the CP stop codon and before the ORF of interest (Fig. 1B). C-terminal CP fusion proteins typically accumulate to high levels in transfected plants. This arrangement allows translational read-through of the amber stop codon at about a frequency of 5%. For some epitopes this 5% read-through product is not significantly excluded from virion assembly and recombinant peptides can potentially be purified as fusion proteins incorporated into soluble TMV virions (Fig. 2A; Turpen *et al.*, 1995). At the other extreme some CP fusions appear to self-aggregate but are not incorporated into virion particles and are recovered as inclusions in low speed pellets. Maintenance of the native RNA conformation at the 3' end of the CP ORF minimizes negative effects on virus amplification (Turpen *et al.*, 1997).

Dual subgenomic vectors

Fusion to the TMV CP will not be desirable for some peptides or larger proteins, including those requiring post-translational modification for active and/or stable conformations. A viral vector containing an additional subgenomic promoter can generally express such proteins. These vectors contain the 5' portion of TMV (including the 126/183 kDa and movement protein genes) followed by unique restriction site(s) after the native CP subgenomic promoter. Foreign genes are inserted into the restriction site(s) immediately downstream of the TMV CP subgenomic promoter (Fig. 1C). The 3' vector portion contains a heterologous tobamovirus coat protein subgenomic promoter, coat protein ORF and 3' non-translated region (NTR; Fig. 1C). This design avoids the duplication of TMV sequences, thereby reducing the frequency of homologous recombination resulting in loss of foreign sequence.

Because genes inserted under the control of the TMV CP subgenomic promoter are not fused to a viral gene, the protein of interest can be directed to various subcellular compartments by the inclusion of specific targeting sequences. For example, functional signal peptide sequences mediate entry into the endoplasmic reticulum (ER). Appropriate sorting information will then localize proteins to the plasmalemma, cell wall, vacuole, tonoplast or back to the ER. In the absence of this sorting information, the protein is secreted to the apoplast. Peptide sequences that direct proteins from the cytosol to the nucleus, mitochondria, chloroplasts, or peroxisomes have also been described (Kermode, 1996) and can be incorporated into foreign polypeptides, if desired.

The subcellular targeting of foreign proteins is generally critical for the accumulation of functional products. When monitoring expression of foreign proteins in plants, one must be careful to measure accumulation and activity in all fractions of the plant extract (Fig. 2) to identify fractions from which to purify and characterize the product. The expression of the human lysosomal enzyme, alpha-galactosidase A (α-Gal A), in plants illustrates the significance of the relationship between subcellular compartmentalization and biological activity of a foreign protein (Fig. 2B, Dawson, 1992).

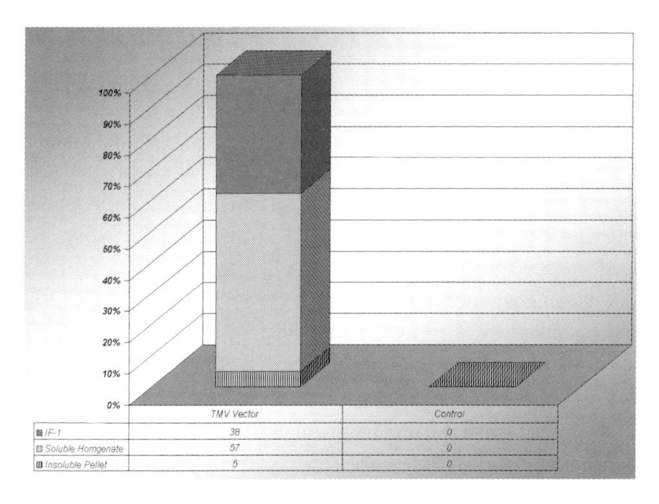

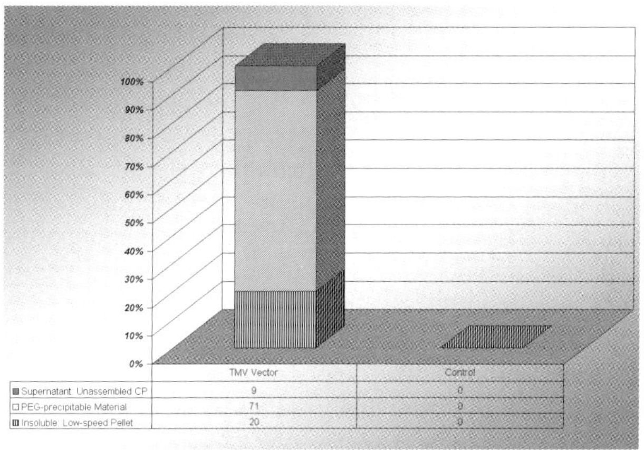

Figure 2. *Generalized schematic of the partitioning of recombinant cross-reacting immunologic material (CRIM) in assembled virions (A) and secreted proteins (B). Key provided for each panel. Ideally during analysis of recombinant protein expression one should account for all CRIM and activity present in the leaf sample. For example, virus particles are convenient purification carriers, provided that a significant portion of total CRIM is recovered in the particulate fraction and not lost in either low-speed pellets (as inclusions) or the polyethylene glycol (PEG) supernatant (as unassembled coat-fusions). Likewise, while the plant leaf does not typically secrete large quantities of proteins to the intercellular fluid (IF), it has a high capacity to do so through the default pathway of protein sorting that relies on bulk membrane vesicle trafficking. Whether or not to use the IF as starting extract depends on the enrichment and yield of protein in this fraction.*

SPECIFIC EXAMPLES OF PROTEIN PRODUCTION USING THE TMV VECTOR

While many proteins have been expressed in TMV vectors, there are some proteins that have been expressed that exemplify the capabilities of the viral vector system. The jellyfish green fluorescent protein (GFP), which fluoresces green under ultraviolet illumination, is very useful for visually showing the speed and expression level of the TMV vector system. In Fig. 3, it is easy to visualize the viral vector expression of the GFP in a temporal way. After 2 days post-inoculation (dpi) with the GFP-containing viral vector, the infection sites and the subsequent localized cell-to-cell movement is observed under UV light. By 4 dpi, the viruses have entered the plant phloem system and are passively transported to points of exit in sink tissues. At this point, the viruses actively leave the vascular system and invade surrounding tissues, rapidly encompassing entire leaves. At 6 dpi, it is easily seen that even the newly emerging leaves are infected by the vector expressing the foreign GFP gene.

Human α-Gal A was expressed in plant leaves with a tobamovirus vector because of the need for an inexpensive source of enzyme to develop enzyme replacement therapy for a hereditary storage disorder known as Fabry disease (Desnick *et al.*, 1995; Pogue *et al.*, 1997; Turpen *et al.*, 1997). The native human cDNA also provided a challenging opportunity to assess the capacity of viral-transfected leaves to accumulate active products because the mature enzyme requires disulfide-mediated folding, N-linked glycan site occupancy, and homodimerization of ~50 kD subunits for activity. Several constructs containing active signal-peptide sequences, for entry into the endoplasmic reticulum, were analyzed by Western analysis for accumulation of cross reacting immunologic material (CRIM) of recombinant α-Gal A (rGal A). Although rGal A accumulated to approximately 2% of total soluble protein in infected leaves, only nominal activity was measured from tissue infected with some constructs. Only after deletion of a portion of a putative carboxy-terminal propeptide was significant enzymatic activity obtained (Fig. 4A,B).

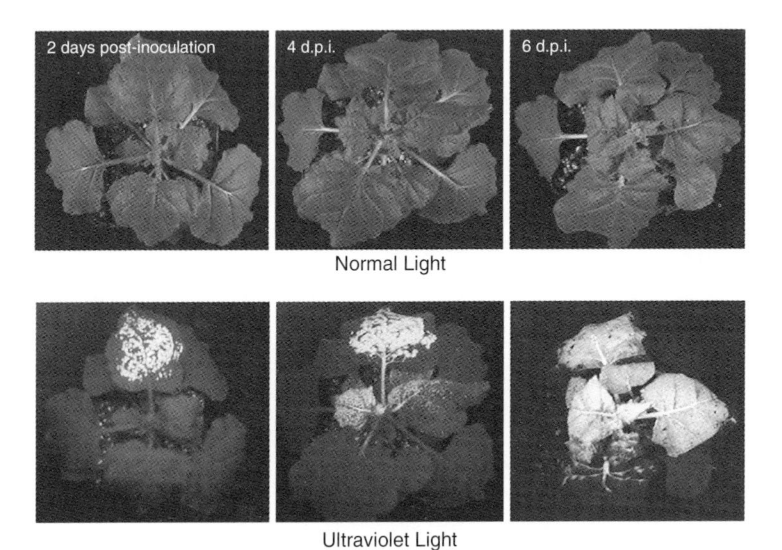

Figure 3. *Time lapse photographs showing the spread of the GFP-containing viral vector. The top three photographs show the same Nicotiana benthamiana plant at 2, 4 and 6 days post-inoculation with the above vector, under normal light conditions. The bottom three photographs show the same plants at 2, 4 and 6 days post-inoculation under ultra-violet light, which allows visualization of the fluorescing GFP. The fluorescing GFP is the result of the spread and expression of the viral vector.*

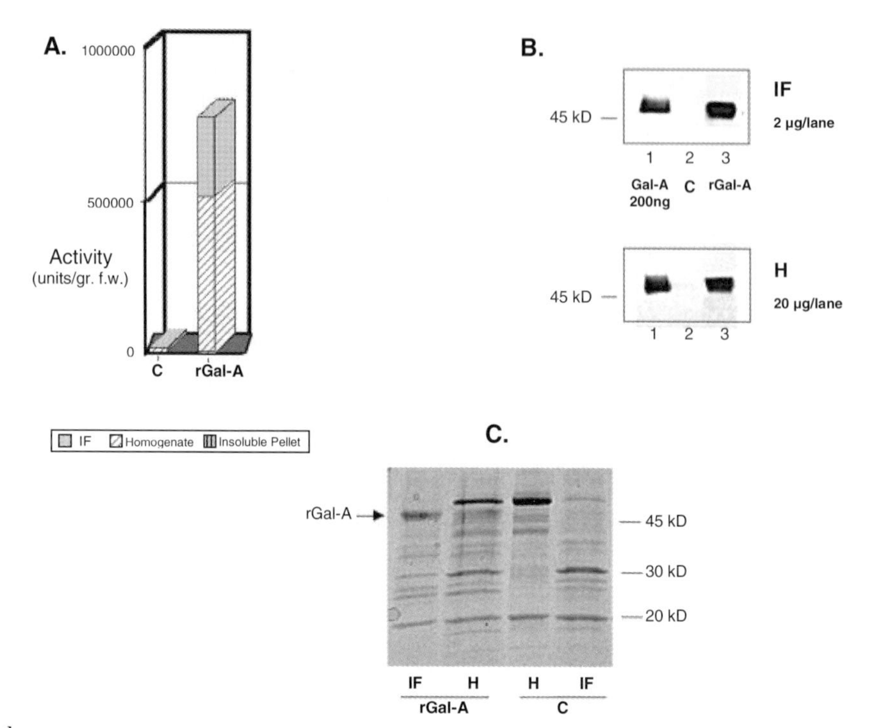

Figure 4. *Quantitative analysis of recombinant α-galactosidase A (rGal A) expression. (A) Partitioning of rGal A enzymatic activity from N. benthamiana leaves infected with control virus (C) and TMV-based vector expressing rGal A: IF-1 (stippled region), homogenate (region with diagonal lines) and insoluble material (region with vertical lines). (B) Western analysis of IF-1 (IF) or homogenate (H) protein from leaf tissue infected with TMV vector expressing rGal-A (lanes 3) and control infected material (C, lanes 2) compared with human enzyme purified from placenta (hGal A; lanes 1). Soluble protein extracts from the IF-1 (2 µg/lane) and tissue homogenates (20 µg/lane) were analyzed by SDS-PAGE and Western blotting using an antibody specific for hGal A. (C) Coomassie staining of protein extracts analyzed in (B). IF-1 (IF) and homogenates (H) from leaf tissue infected with control virus and TMV vector expressing rGal A were separated by SDS-PAGE. Sizes of molecular markers are identified to the right and strongly staining bands corresponding to rGal A in IF-1 and homogenate are indicated by arrow to the left of panel. Equivalent protein amounts from the IF and homogenates are loaded in each analysis.*

This activity was recovered from the plant apoplast, a network of intercellular fluid (IF), cell wall and extracellular matrix material located between adjacent cells in leaves. Because leaf cells do not normally secrete large quantities of protein, rGal A was by far the predominant species (>30%) of protein recovered in the IF (Fig. 4C). The enzyme purified from the leaf IF has a specific activity equal to or greater than that of the enzyme purified from human tissues and other recombinant sources.

While lower-cost production of complex, therapeutic proteins is an ideal opportunity for viral vectors in plants, the rapid expression capabilities of the viral vector can provide the opportunity to produce "quick turnaround" patient-specific therapeutics. For example, the TMV vector has been used to rapidly produce a single-chain Fv fragment (scFv) that was successfully used as a therapeutic vaccine in a mouse model for B-cell lymphoma (McCormick *et al.*, 1999). Within 4 weeks of preparing and cloning the DNA encoding the scFv protein, the sequence was inserted into the dual subgenomic viral vector and the transfected plant tissue was expressing the desired protein. The design of the scFv protein included a rice α-amylase signal peptide that targeted the protein into the secretory pathway (for proper folding). The scFv could then be easily purified from the apoplast of infected plants (McCormick *et al.*, 1999). The capability of the viral vector to produce functional scFv proteins in a matter of weeks could likely enable treatment of human lymphomas, as well as other cancers, through patient-specific immunotherapy.

CURRENT LIMITATIONS OF THE VIRAL VECTOR SYSTEM

While the viral vector system has speed and yield as primary traits, there have been some limitations that have yet to be overcome. One such limitation is that the current viral vectors have not yet been able to express proteins in seed tissue. As such, green tissue and roots are the primary source of the viral vector-produced proteins. Freshly harvested green tissue cannot be stored for long periods, as can seeds, and must be extracted relatively quickly after harvest. However, it may be possible to dry the green tissue and extract the peptides or protein at a later date without a significant loss of product. Permanent traits are not possible using the current vectors, as they are unable to go through pollen or seed to subsequent generations, and are therefore limited to one generation of the plant. Another limitation is the size constraint of the genes that the viral vectors can successfully produce in high yields. Currently, yield and stability of the tobamoviral vector become problematic when expressing non-viral, foreign proteins that exceed approximately 100 kDa. However, it may be possible for viral vectors derived from different viruses to improve on the current size constraints. Finally, current viral vectors have reduced "fitness" when carrying additional genes that are not needed for virus replication. In most cases, the foreign gene will eventually be lost from the virus. However, for short-term production of proteins and peptides, this is not a limitation. In fact, the eventual loss of the foreign gene insert into viral vectors may be a beneficial attribute, especially in light of the current GMO concerns related to transgenic plants.

SUMMARY

For over 10 years, TMV-based vectors have been used as plant expression tools to examine gene regulation and function, protein processing, pathogen elicitors, to manipulate biosynthetic pathways, and to produce high levels of enzymes, proteins, or peptides of interest in different locations in a plant cell. TMV vectors often exhibit genetic stability of foreign RNA sequences through multiple passages in plant hosts. Foreign coding sequences can be expressed in plants where the stability, intracellular fate and enzymatic or biological activities of the recombinant proteins can be rapidly evaluated and optimized. These properties make viral vectors attractive expression vehicles for testing and production of a wide variety of recombinant peptides and proteins, for structural analyses of post-translational modifications and for assessing gene function and metabolic control. Finally, the utility of both CP fusion and dual subgenomic vectors has extended beyond the laboratory and greenhouse to field-scale production and purification of recombinant products for commercial use (Grill, 1992; Grill, 1993; Turpen *et al.*, 1997).

REFERENCES

Copeman RJ, Hartman JR and Watterson JC. 1969. Tobacco mosaic virus in inoculated and systemically infected tobacco leaves. *Phytopathology* 59:1012-1013.

Dawson WO, Beck DL, Knorr DA and Grantham GL. 1986. cDNA cloning of the complete genome of tobacco mosaic virus and production of infectious transcripts. *Proc. Natl. Acad. Sci.* (USA) 83:1832-1836.

Dawson WO and Lehto KM. 1990. Regulation of tobamovirus gene expression. *Ad. Virus Res.* 38:307-342.

Dawson WO. 1992. Tobamovirus-Plant Interactions. *Virology* 186:359-367.

Deom CM, Lapiodot M and Beachy RN. 1992. Plant virus movement proteins. *Cell* 69:221-224.

Desnick RJ, Joannou YA and Eng CM. 1995. α-Galactosidase A Deficiency: Fabry Disease, In: The Metabolic Bases of Inherited Diseases, C.R. Scriver, A.L. Beaudet, W.S. Sly, and D. Valle (eds.) McGraw-Hill, pp. 2741-2784.

Donson J, Kearney CM, Hilf ME and Dawson WO. 1991. Systemic expression of bacterial gene by a tobacco mosaic virus-based vector. *Proc. Natl. Acad. Sci.* (USA) 88:7204-7208.

Fitchen J, Beachy RN and Hein MB. 1995. Plant virus expressing hybrid coat protein with added murine epitope elicits autoantibody response. *Vaccine* 13:1051-1057.

Goldbach R and Holn T. 1997. Plant viruses as gene vectors. *Meth. Plant. Biochem.* 10b:103-120.

Gooding GV and Herbert TT. 1967. A simple technique for purification of tobacco mosaic virus in large quantities. *Phytopathology* 57:1285.

Grill LK. 1992. 1991 Tobacco field trials report and soil and plant analysis follow-up on the 1991 tobacco field trials report; Filed with the USDA-APHIS, Hyattsville, MD.

Grill LK. 1993. Tobacco mosaic virus as a gene expression vector. *Agro. Food Industry Hi Tech.* Nov/Dec. 20-23.

Hamamoto H, Sugiyama Y, Nadagawa N, Hashida E, Matsunaga Y, Takemoto S, Watanabe Y and Okada Y. 1993. A new tobacco mosaic virusvector and its use for the systemic production of angiotensin-I-converting enzyme inhibitor in transgenic tobacco and tomato. *Bio/Techology* 11:930-932.

Kearney CM, Donson J, Jones GE and Dawson WO. 1993. Low level of genetic drift in foreign sequences replicating in an RNA virus in plants. *Virology* 192:11-17.

Kermode AR. 1996. Mechanisms of intracellular protein transport and targeting in plant cells. *Crit Rev. Plant Sc.* 15:285-423.

Kumagai MH, Turpen TH, Weinzettl N, Della-Dioppa G, Turpen AM, Donson J, Hilf ME, Grantham GL, Dawson WO, Chow TP, Piatak, Jr M and Grill LK. 1993. Rapid, high-level expression of biologically active α-trichosanthin in transfected plants by an RNA viral vector. *Proc. Natl. Acad. Sci.* (USA) 90:427-430.

Kumagai MH, Donson G, Della-Cioppa G, Harvey D, Hanley K and Grill LK. 1995. Cytoplasmic inhibition of carotenoid biosynthesis with virus-derived RNA. *Proc. Natl. Acad. Sci.* (USA) 92:1679-1683.

Kumagai MH, Donson J, Della-Cioppa G and Grill LK. 2000. Rapid, high-level expression of glycosylated rice α-amylase in transfected plants by an RNA viral vector. *Gene* 245:169-174.

McCormick AA, Kumagai MH, Hanley K, Turpen TH, Hakim I, Grill LK, Tus D, Levy S and Levy R. 1999. Rapid Production of Specific Vaccines for Lymphoma by Expression of the Tumor-derived single-chain Fv Epitopes in Tobacco Plants. *Proc. Natl. Acad. Sci.* (USA) 96:703-708.

Namba K, Pattanayek R and Stubbs G. 1989. Visualization of protein-nucleic acid interactions in a virus: refined structure of intact tobacco mosaic virus at 2.9 A resolution by X-ray fiber diffraction. *J. Mol. Biol.* 208:307-325.

Pogue GP, Turpen TH, Hidalgo J, Cameron TI, Murray GJ, Brady RO and Grill LK. 1997. Production and purification of a highly active human α-galactosidase A using a plant virus expression system. Abstract. *Amer. Soc. Virol. Meeting*. Bozeman, MT; p 162

Scholthof HB, Scholthof K-BG and Jackson AO. Jackson 1996. Plant virus gene vectors for transient expression of foreign proteins in plants. *Annu. Rev. Phytopathol.* 34:299-323.

Siegal A, Hari V and Kolacz K. 1978. The effect of tobacco mosaic virus infection on host and virus specific protein synthesis. *Virology* 85:494-503.

Sugiyama Y, Hamamoto H, Takemoto S, Watanabe Y and Okada Y.H. 1995. Systemic production of foreign peptides on the particle surface of tobacco mosaic virus. *FEBS Let.* 359:247-250.

Turpenc TH, Reinl SJ, Charoenvit Y, Hoffman SL, Fallarme V and Grill LK. 1995. Malarial epitopes expressed on the surface of recombinant tobacco mosaic virus. *Bio/Technology* 13:53-57.

Turpen TH, Cameron TI, Reinl SJ, Pogue GP, Garger SJ, McCulloch MJ, Holtz RB and Grill LK. 1997. Production of recombinant proteins in plants: Pharmaceutical applications. The Soc. Exper. Biol., Canterbury, U.K. *J. Exp. Botany* (Suppl.) 48:12.

Zaitlin M and Israel HW. 1975. Tobacco mosaic virus (type strain). C.M.I./A.A.B. Descriptions of Plant Viruses. Wm. Culross and Son, Ltd. UK.

ALFALFA, A PERENNIAL SOURCE OF RECOMBINANT PROTEINS

Louis-Philippe Vézina[1], Marc-André D'Aoust[1], Damien Levesque[2], Patrice Lerouge[3], Véronique Gomord[3], Loïc Faye[3], Mark McCaslin[4], François Arcand[1]

[1] Medicago inc., Ste-Foy, Québec, Canada G1K 7P4
[2] Avenir Luzerne, Aulnay-les-Planches, France
[3]CNRS, Laboratoire des transports intracellulaires, CNRS, Université de Rouen, Mont Saint-Aignan cédex, France
[4]Forage Genetics International, West Salem, WI USA

INTRODUCTION

Alfalfa (*Medicago sativa* L.), the most important forage crop in the world, is now finding a new mission as a bioreactor for the production of useful proteins. Its advantageous agronomic characteristics, as well as its simple composition with regards to secondary metabolites, has motivated its choice as cellular factory for protein production. Alfalfa is a perennial crop that will yield 10 to 12 tons of dry matter for up to 6 years in the field, and more than 10 years when grown in a greenhouse. It fixes atmospheric nitrogen through a symbiotic association with root-invading bacteria, and thus has no fertilization requirements, if not for microelement deficiencies in some soil types. The plant is also propagated easily by stem cuttings, and has a strong regenerative capacity, which gives the possibility to produce large clonal populations within a short time frame. Finally, purification of recombinant proteins from its sub-fractions is fast and easy duc to past industrial developments on wet fractionation processes for animal feed production.

It was the drastic change brought by globalization of agriculture and its threat for the economics of the dairy industry that triggered an intensification of research for new uses for alfalfa, one of which being its use as cellular factory. Proof is the mother of certainty; and following the first signs of changes in the economics of the dairy industry, some institutions once devoted to research on forage improvement, became interested in developing the first applications of alfalfa for molecular farming. The crop has now proven its potential as a source biomass for the production of high valued recombinant proteins. This chapter will summarize the work that has been done at the academic and commercial levels, which resulted in the recent development of industrial-scale processes for the production of purified recombinant compounds from alfalfa.

E.E. Hood and J.A. Howard (eds.), Plants as Factories for Protein Production, 17–54.

Molecular farming with plants is a five-step process through which a cDNA encoding for a protein of interest is first incorporated in an expression cassette (1), transferred to a plant cell (2), expressed as a "self" protein by the growing transgenic organism (3), partially or totally extracted and purified from the organism (4) while the co-products of extraction are destroyed or processed in ways that insure bio-safety and optimize profitability (5).

Molecular farming must therefore be seen first as an industrial process that comes as an alternative to the production of recombinant proteins with yeast, bacteria or mammalian cells. While scientific comparisons about production levels per mass unit, proper folding or potential allergenicity of translation products often seem central to the debate between these different production systems, it is within the frame of profitability, public concern, scale-up strategies, bio-safety, and world economics that molecular farming will demonstrate its true potential.

Molecular farming currently goes through an iterative development, which is comparable to that through which most currently used high technologies went, and it will eventually become part of the modern life-style. As an example, from the time of its birth and successful testing as a carrier for spacecrafts, the Saturn propeller, underwent more than 7,000 major modifications before it was launched, assembled to Apollo and LEM, for the first human odyssey to the moon. On this memorable day, when it released its enormous thrust to the minuscule spacecraft bolted to its top, the Saturn propeller had been passed from the hands of its creators, to those of hundreds of second-generation parents which participated in its final shaping. Similarly, if molecular farming is to become a reliable, globally accepted and safe method to produce the thousands of new therapeutics that promise to change the current perception of health in the next decades, it has to face the constraints and dynamics of world economics, financing, environmental concerns, and most of all, profitability; its science and technologies now undergo the critical modifications that are crucial to the birth of all revolutionary industries.

We have shaped this chapter around the five basic industrial activities required to perform molecular farming as a business; this was intentional not to diminish the impact of technological improvements or scientific breakthroughs in the realm of molecular farming, but to insist on putting these recent findings in the context of an industrial activity. Each of these steps will be described in the perspective of their relative importance within the process as a whole, with special emphasis when appropriate, on underlying science, intellectual property and large-scale feasability. In the introduction to this chapter, the biology and agronomy of alfalfa are presented that should help understanding the advantages of the alfalfa system for the production of recombinant proteins.

THE BIOLOGY OF ALFALFA

Morphology, Anatomy and Physiology

Alfalfa has small kidney shaped seeds. The average seed count is 464 seeds per gram (Teuber and Brick, 1988). Its seed color ranges from yellow to olive green. Impermeable or "hard seeds" are common when seed is produced in northern areas. These seeds fail to immediately imbibe water when placed in a moist environment. However, most hard seed will germinate and successfully establish within 30 days of seeding (Undersander, unpublished). Upon germination the radical elongates rapidly and penetrates the soil as an unbranched tap root. As the hypocotyl elongates, the cotyledonary leaves emerge from the soil. The primary shoot continues to elongate as the seedling grows.

Alfalfa is primarily tap-rooted, but there is considerable genetic variation for root growth habit. Alfalfa is deep-rooted, commonly rooting to a depth of 5 meters or more. This helps give the crop excellent drought tolerance. The deep-rooted nature of the crop, combined with the biological activity of the alfalfa rhizosphere help improve soil tilth and soil fertility making alfalfa an excellent crop in rotation with other crop species.

Concurrent with early root development and elongation of the primary shoot, the hypocotyl region undergoes contractile growth, pulling the cotyledons and developing crown below the soil surface. This contractile growth begins 2-6 weeks after germination and helps insulate crown buds from adverse environmental conditions (e.g. extremes in temperature). The primary crown develops from the axils of the unifoliate leaf and the cotyledons. Further crown development occurs later from nodes in the proximity of the primary crown and on nodes of primary shoots below cutting height. The alfalfa crown is a complex structure, with both genetic and environmental variation affecting size and depth of the crown below the soil surface. The crown is the essential feature of perenniality in alfalfa. Crown buds are the sole source of new growth during spring green up and a primary source of new growth after each cutting.

Alfalfa is a perennial forage crop with stands generally lasting 3-6 years. Plants become dormant in the fall and green up the following spring. Fall dormancy is triggered by both decreasing day length and lower temperatures. Cultivars vary widely in their day length sensitivity (i.e. fall dormancy). Spring green up occurs when crown buds, formed the previous season, begin growth in response to longer days and warm spring temperatures. Timing of green up also depends on plant health and genetic fall dormancy (sensitivity to day length). Sub-lethal winter injury often results in slow and uneven spring recovery.

Alfalfa is harvested multiple times during the growing season. Regrowth after cutting arises from crown buds and axillary buds on the lower stem, below cutting height. Ideal cutting height is 2 inches above the soil level. This provides an optimal number of axillary buds for abundant regrowth potential. Short cutting intervals, biotic and abiotic stress may decrease the number of axillary and crown buds, resulting in a decrease in stem density and forage yield of the succeeding crop.

The shorter cutting intervals used for the production of high quality forage may also negatively impact persistence. During the cutting interval storage carbohydrates and storage proteins are accumulated in the crown and root of alfalfa. These reserves are used to fuel both spring growth and regrowth after cutting. Short cutting intervals may not allow the full replenishment of carbohydrate and protein reserves and lead to higher risk of winter injury and decreased persistence.

The leaves of alfalfa are arranged alternately and are primarily trifoliate, though multifoliolate types have been identified and used extensively in breeding. Alfalfa leaves are more digestible and have more nutrients than stems. Leaves generally contain two to three times as much crude protein than alfalfa stems (Mowat, et. al., 1965). Leaves also lose their quality more slowly with advanced maturity (Kilcher and Heinrichs, 1974). Leaves generally make up about 50% of the harvested dry matter. The leaf to stem ratio decreases as the crop matures. This, along with the lignification of secondary cell walls in older stems, contributes to the overall loss in alfalfa forage quality with advancing maturity.

Alfalfa flowers grow in clusters (racemes) attached to the stem. Most alfalfa cultivars have primarily purple flowers. Purple flowers are typical of germplasm tracing solely or predominantly to *Medicago sativa* spp *sativa*, with a center of origin in Iran/Turkestan. *Medicago sativa* spp *falcata*, with a center of origin in the dry Steppes of southern Russia has yellow flowers. Many winterhardy alfalfa cultivars contain germplasm from both sources and have a combination of purple, yellow and variegated flower color.

Alfalfa flowers are insect pollinated and most plants have low self-fertility. Seeds are formed in a coiled fruit (pod), with each pod containing 5-15 seeds. Seeds generally mature 5-6 weeks after pollination.

Genetics

Cultivated alfalfa is an autotetraploid (2n=16) perennial. The autotetraploid nature of the crop makes inheritance quite complex. Breeding systems have been developed to facilitate the incorporation of both conventional and transgenic (McCaslin, unpublished) traits into autotetraploid alfalfa cultivars are heterogeneous populations. The large genetic variation within cultivars generally gives them broad adaptation, but can also present breeding challenges.

Alfalfa is very sensitive to inbreeding depression. Losses in forage yield and seed yield are associated with even moderate inbreeding. The autotetraploid nature of the crop creates multiple levels of heterozygosity. Two, three or four independent alleles at a single locus give three levels of heterozygosity: diallelic, triallelic and tetrallelic, respectively. Plant breeding programs should be structured to minimize inbreeding (Hill, 1976) and maximize heterozygosity (Bingham, 1979) in alfalfa cultivars.

Alfalfa is successfully grown from inside the artic circle to the arid deserts of the Middle East. Ecotypes have been identified that tolerate extremes in both temperature and moisture. The enormous natural genetic variability within *Medicago* and related species has been a rich source of genetic variation for alfalfa breeders. Genes for resistance to virtually all the major disease and insect pests in alfalfa have been isolated from this natural gene pool.

THE AGRONOMY OF ALFALFA

Establishment and fertility

Successful stand establishment requires the right conditions for germination. A firm seedbed insures good seed to soil contact and is critical for good establishment. Proper seeding depth is 0.2 to 0.5 inch in medium to heavy soils and 0.5 to 1 inch in sandy soils. This seeding depth insures moist conditions for germination while allowing the elongating hypocotyl to reach the soil surface.

The optimum soil pH for alfalfa production is 6.5 to 7.5. As pH falls below the optimum, forage yield is significantly affected. In a 1984 study of soils in the US Midwest, a decrease in soil pH from 6.6 to 6.0 resulted in a 7% loss in forage yield (McLean and Brown, 1984). Poor Rhizobia nodulation of alfalfa is common at low soil pH. Low soil pH is also often associated with Al and/or Mn toxicity in alfalfa. The solubility of both chemicals is increased at low soil pH. In general, nutrient needs will depend on other management practices and fertilizer use must be consistent with the overall level of management.

Alfalfa is harvested multiple times during the growing season, with the forage generally fed as hay, haylage or green chop. The number of harvests and the interval between cuttings varies widely based on climate/geography, availability of moisture and nutrients, cultivar, and management. In southern California, alfalfa is harvested up to ten times per year over an eleven-month season, four times a year in Eastern France and twice a year in North Dakota and northern Quebec.

Cutting management is an important tool in achieving high quality, high yields, and stand persistence. It can also be effective in reducing the impact of weed, insect and disease pests. From a developmental standpoint, forage yield and forage qualities are inversely related. Harvest of alfalfa in the vegetative stage (i.e. before flower development) will result in high forage quality and low forage yield. Conversely, delayed harvest results in increased forage yield and decreased forage quality. This dilemma has implications for both the dairy farmer and the molecular farming enterprise. High forage quality for the dairy producer means high protein content, high forage digestibility and high animal intake potential. High quality for the molecular farming enterprise will be the content of the recombinant protein in the forage. In both cases the goal will likely be a cutting management system that maximizes the yield of digestible nutrients, or recombinant protein, per acre.

As noted earlier, cutting management can also significantly affect persistence. In general, the less aggressive the cutting management (i.e. longer cutting intervals), the longer the stands will last. Stand longevity is also significantly affected by weather, genetics and soil fertility, primarily potassium.

Alfalfa Varieties

Alfalfa cultivars have been developed with specific adaptation to all major alfalfa production areas. Fall dormancy response is a primary criterion for determining such adaptation. Fall dormancy is a measure of day length sensitivity and is rated on a 1-10 scale based on height of fall growth one month after an early September harvest. Fall dormancy 1 (FD1) types exhibit virtually no growth after the September harvest and actually show much decreased vegetative growth as early as mid-August.

Table 1. *General areas of adaptation for various fall dormancy types in alfalfa*

FD class	North America	South America	Europe	Other
FD2-4	Northern U.S. and Canada		Poland and Russia	Northern China
FD4-6	U.S. Central Plains and Southeast	Southern Argentina and Chile	France, Denmark, Germany, and Hungary	Japan
FD6-7	Central California	Argentina and Chile	Italy and Spain	Australia
FD8-9	Southern CA and Mexico	Argentina, Chile and Peru	Greece	Middle East, India, Australia and South Africa
FD9-10	Southern CA and Mexico	Northern Argentina		Middle East

FD10 types are essentially day length insensitive and show winter active growth in areas where winter temperatures are mild. Table 1 shows typical areas where the various fall dormancy types are grown.

Alfalfa varieties are typically placed in one of three general fall dormancy groups: Dormant (FD2-5), Semi-dormant (FD6-7) and Non-dormant (FD8-10). Typically the longer the available growing season, the less dormant the variety planted. The proper variety allows forage production for the full season available.

Winter hardiness is the ability to survive cold temperature stress during the winter and early spring. Although there is a general negative relationship between fall dormancy and winter hardiness (i.e dormant varieties are more winter hardy than non dormant varieties), there is considerable genetic variation for winter hardiness within a dormancy class. Winter injury is a major limiting factor in alfalfa production in most northern climates where alfalfa is grown. Winter hardiness is a second major factor determining adaptation for alfalfa cultivars.

A third variable affecting adaptation is resistance to specific diseases and insects. Alfalfa breeders have developed germplasm sources with resistance to virtually all of the major disease and insect pests in alfalfa. Each major alfalfa production area has at least one or two pests that can significantly reduce alfalfa yield or persistence. Using this information, specific pest resistance requirements can be developed for each growing region. Linked with requirements for winter hardiness and fall dormancy, specific pest resistance requirements can be used to craft varieties with specific regional adaptation.

There is also considerable genetic variation for forage quality potential (e.g. increased digestibility), tolerance to intensive grazing, improved tolerance to saline soils, and increased drought tolerance that may affect variety choice.

This dilemma has implications for the molecular farming enterprise. Molecular Farming is often seen as focused on the highest valued product it generates, the recombinant protein. Realistic cost structure schemes emphasize the fact that abundance and variety of co-products are of importance in the overall acceptance and economics of molecular farming. For this reason, care is taken to manage variety choice, growth and harvest techniques so that abundance and quality of co-products are maintained.

TRANSGENESIS

Expression cassettes

As for most other crops, the initial genetic transformation work that was done with alfalfa was with the use of commercially available expression cassettes. Most of these cassettes contained a functional assembly of the cauliflower mosaic virus (CaMV) 35S promoter (Odell *et al.*, 1985; Jefferson *et al.*, 1987) or its derivatives and a transgene of choice. The CaMV 35S promoter drives a relatively weak transcription wave in alfalfa. A comparative study has shown that steady-state messenger RNA levels for the tomato proteinase inhibitor I under the transcriptional control of 35S is 3 to 4 times lower in alfalfa than in tobacco (Narváez-Vásquez *et al.*, 1992). This and other factors such as high ploidy level, restrained the efforts
in developing alfalfa as a cell-factory. However, results obtained with the over-expression of a monoclonal antibody (Khoudi *et al.*, 1999) and the SFA8 (Tabe *et al.*, 1995) protein of sunflower revived the belief that high-yielding perennial plants could provide an interesting alternative to annual seed crops. Following is a summary of the recent improvements in the development of efficient expression cassettes for alfalfa.

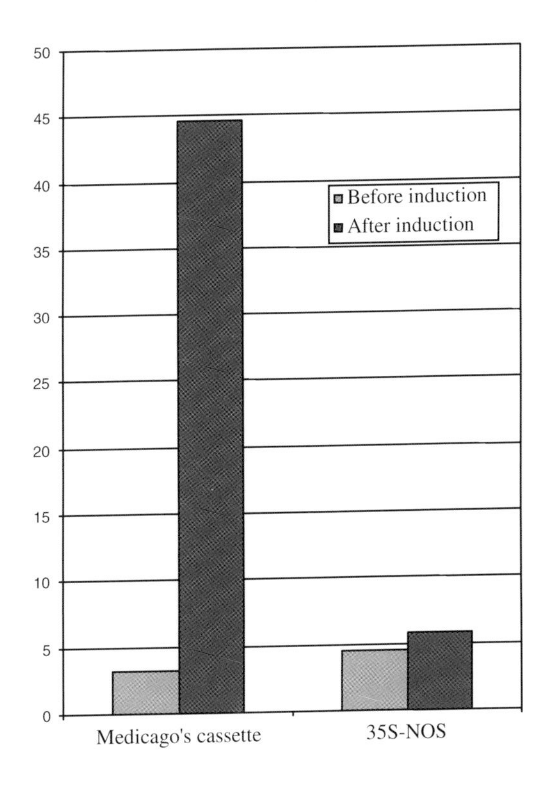

Figure 1. *Inducibility of GUS expression with Med-1302T expression cassette.*

New developments

Poor yields with currently available cassettes, and intellectual property issues, have led to the development of new strong promoters suitable for expression in alfalfa. In one study, the use of the alfalfa Rubisco small subunit (RbcSK-1A) promoter resulted in expression of the GUS gene with more plant-to-plant variability than with the 35S promoter. Despite variations, mean expression levels obtained with the RbcSK-1A promoter were 10-fold that of the 35S promoter (Khoudi *et al.*, 1997). We have pushed forward the identification of strong promoters for inducible and constitutive expression in alfalfa. Two main parameters have directed the efforts in this development:

biosafety and yield. With current concerns regarding the co-introduction of undesirable DNA upon cell transformation, it seemed desirable to develop promoters and expression cassettes, which consist of alfalfa regulatory fragments, and thus would not raise the question of transspeciation. As with most heterologous expression systems, optimal yields are not obtained with constitutive expression, notwithstanding the potential undesirable affects of constitutive expression on the plant itself, and its environment. We have developed a series of cassettes, which contain a new inducible promoter derived from alfalfa regulatory sequences. With this promoter, expression is induced by a safe inorganic compound (patent pending). In Fig. 1, we compare the efficiency of this new promoter with that of the 35S promoter using β-glucuronidase (GUS) reporter gene expression in tobacco. Once induced, this proprietary promoter is ten-fold stronger than the 35S promoter. Recent results with genuine constructs expressed in alfalfa show similar induction levels but no GUS activity can be detected before induction.

Following the same general principles, three families of cassettes have been developed from alfalfa regulatory elements for expression in alfalfa foliage; some are light inducible and others show inducibility with developmental stages of the leaves. Among these, a series of light inducible cassettes (patent pending) direct 25-times higher expression of the GUS reporter gene than the 35S promoter does.

Intra-cellular targeting

There are only a few referenced studies that exemplify the need to target protein accumulation in alfalfa, one of which, by Wandelt and co-workers (1992), demonstrated a 20-fold increase in Vicilin (a vacuolar pea seed storage protein) accumulation when the protein was retained in the endoplasmic reticulum (ER) by the addition of a carboxy-terminal KDEL sequence to the protein. Wandelt and coworkers showed that this accumulation was due to an increased stability of the protein *in planta*. Later, it was demonstrated that the accumulation of a sunflower seed storage albumin (SFA8) in alfalfa leaves was dependent on the addition of an ER retention signal (Tabe *et al*., 1995). No SFA8 protein could be detected in the plants harboring the SFA8 gene without ER retention signal, even if the corresponding mRNA was abundant. The authors concluded that the ER retention signal kept the protein from going to the highly proteolytic leaf vacuole, the vacuole being the natural accumulation organelle of seed storage proteins like SFA8. Bagga and coworkers (1992) showed evidence for another example of vacuolar instability with recombinant proteins in alfalfa leaves. When expressing the ß-phaseolin protein under the control of the 35S promoter, they found no ß-phaseolin accumulation in the leaf tissue. Interestingly, it was shown that the seeds of the transgenic plants accumulated ß-phaseolin. Similar results were described by Benmoussa *et al* (1999) who over-expressed the gene encoding glutenin, a wheat vacuolar protein, in alfalfa: the protein was only found in the seeds of transgenic plants.

However few, these reports show that alfalfa leaves are a highly active tissue, in which proteolysis is part of programmed rapid protein turn-over, in comparison to specialized and more quiescent tissues such as seeds in which proteolysis is partly interrupted by dehydration. They also show that proteolysis can be counteracted by proper targeting.

IN VITRO CULTURE

Somatic embryogenesis

Despite the facility with which alfalfa produces roots and new stems from cuttings, early studies demonstrated that alfalfa does not undergo organogenesis in vitro. Pioneering work on alfalfa regeneration through somatic embryogenesis dates to early 1980's (Kao and Michayluk, 1980; Johnson *et al.*, 1981; Groose and Bingham, 1984; Atanassov and Brown, 1984; Brown and Atanassov, 1985). These key experiments indicated that regeneration must proceed through three steps: callus induction, embryogenesis induction, and embryo maturation, each stage being controlled by typical auxin/cytokine ratios. Callus tissue is better obtained from stems, petioles or leaves in modified B5 medium (Chabaud *et al.*, 1988); undifferentiated cells will start dividing within a few days of transplantation of explants in culture medium. Depending on the genotypic source of the explant, embryogenesis will be initiated soon after callus development, or be delayed until different hormonal concentrations are imposed. Typically the callus stage will proceed over 28 days. The explants and calluses can then be transferred to SH medium for embryogenesis induction (Schenk and Hilderbrandt, 1972). Embryos develop from sub-epidermal cells soon after transfer; they will undergo polarized development until emergence from the epiderma, and development will proceed to a typical torpedo-shaped stage, at which embryos can be transferred to embryo maturation medium for a BOi2Y (Blaydes, 1966). Further developments will be accompanied by cotyledon formation, and elongation of the hypocotyl and root axis. Embryos will not set roots in BOi2Y medium. Rooting and stem development will only take place in MS or MS IBA mediums (Murashige and Skoog, 1962). The structures developing during somatic embryogenesis in alfalfa are illustrated in Fig. 2.

Depending on genotypic source, in vitro regeneration of fully developed plants from explants will demand 35 to 70 days. With highly embryogenic material, embryos will readily appear after two weeks in B5H and transferred directly to BOi2Y for maturation. On the contrary, some genotypes are non-embryogenic and will never produce embryos from non-embryogenic tissues.

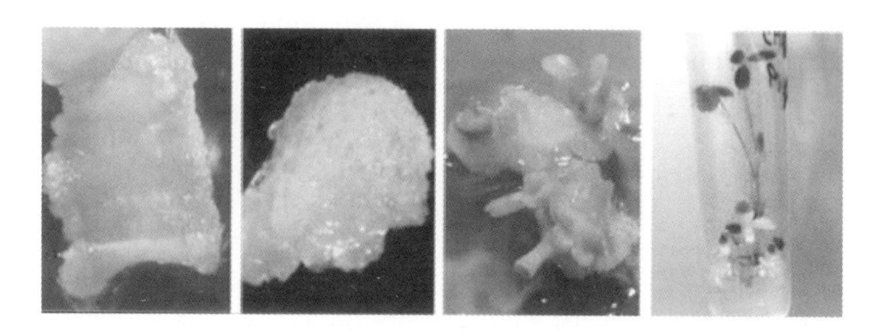

Figure 2. *Somatic embryogenesis in alfalfa. A callus is formed from the division of undifferentiated cells from the explant (1). Some of the cells in the callus differentiate into embryogenic cells (2). The embryos develop and mature (3). The mature embryo is separated from the callus and grows into a fully developed plant (4).*

Suitable protocols for the regeneration of mature plants from protoplast cultures have also been established (Kao and Michayluk, 1980; Johnson *et al.*, 1981; Atanassov and Brown, 1984; Song *et al.*, 1990, Holbrook *et al.*, 1985). To this day, regeneration from protoplasts requires highly regenerative material and skilled in vitro manipulation. It has been demonstrated that walled cells derived from protoplasts are amenable to embryogenesis through processes similar to cells obtained from liquid cultures or from tissue-derived calli. However, until recently, it was thought that isolated protoplasts regenerated at low rates, and most protocols benefited from the early formation of cell clusters. There are no recent reports on regeneration of alfalfa plants from protoplasts albeit the increasing number of reports on the physiology and molecular biology of embryogenesis, and improvements in regeneration protocols for recalcitrant species (see genotype specificity section). Of particular interest are the use of immobilization systems such as agarose or alginate beads, and the addition of specific polyamines and monosaccharides to the culture media.

Genotype specificity

As a result of poor yields in plant regeneration, extensive screening has been performed by different groups to identify and characterize highly embryogenic genotypes (Brown and Atanassov, 1985; Matheson *et al.*, 1990; Nowak *et al.*, 1992). This strategy allowed for the drastic improvement in regeneration frequency, and paved the way to more efficient transformation systems. It was soon established that most of the genetic potential for regeneration came from the *Medicago falcata* background (Brown and Atanassov, 1985). Different trends were followed for the isolation of highly regenerative material. Some groups elected to screen within populations containing high *M. falcata* background, and ended up with some of the most highly regenerative material (Brown and Atanassov, 1985; Wan *et al.*, 1988; Wandelt *et al.*, 1991; Micallef *et al.*, 1995). These genotypes have poor agronomic qualities and transformants obtained from these genotypes must be backcrossed to more suitable material. Other groups performed selection within far larger populations from commercial breeding lines, and ended up with rare regenerative genotypes from which transgenics can now be obtained without the need for backcrosses (Nowak *et al*, 1992; Desgagnés *et al* 1996). Although both types of genotypes are still used for the production of transgenics, as the pressure for short time frame increases, it is believed that production of first transformation events should be performed with as commercially fitted genotypes as possible. Still, the extremely high embryogenic potential of some genotypes with high *M. falcata* background is advantageous in many circumstances. These genotypes will usually produce a larger number of embryos per explant, they will undergo easily embryogenesis in liquid culture and they will produce embryos in a shorter time frame.

Transformation with *Agrobacterium*

Successful transformation of alfalfa through *Agrobacterium* infection dates back to 1986 (Deak *et al.*, 1986; Shahin *et al.*, 1986). *Medicago* was among the first species for which stable transformants were obtained with a bacterial DNA transfer vector. However, unlike for tobacco, regeneration of mature alfalfa plants proceeds through somatic embryogenesis, and thus, efficiency of transformation was for a long time closely related to the ability to develop protocols for in vitro regeneration.

Alfalfa transformation with *Agrobacterium* is easily accomplished by co-cultivation with wounded explants, provided a suitable genotype is used. It was established that regenerative potential of a certain genotype is not the only criterion for suitability to transformation. In a study by Desgagnés and coworkers (1995), it was demonstrated that three highly regenerative genotypes gave widely different yields of transgenics, ranging from 1-2 to 50 independent transformation events per 50 leaf disks (see Table 2). Unpublished results from a screening for transformability in our lab has shown that some highly regenerative genotypes will show very early signs of senescence in the presence of disarmed *Agrobacterium* strains, and thus will never produce transgenics.

Typically, 20 to 25% of leaf or petiole explants will form embryogenic calluses when transformation is performed in permissive conditions; selection of transformants for kanamycin resistance can be performed on 25 mg/L for most genotypes, with no escapes. Polyembryogenesis is the development of secondary embryos on embryogenic tissue and it will frequently occur when one callus undergoes embryo formation. Although these secondary embryos are not independent transformation events and are of no use for the objective of producing transgenic material, this characteristic has been used to develop other transformation strategies (see particle bombardment section below). Co-cultivation with *Agrobacterium* will slow down the regeneration process, but transgenic embryos have frequently been obtained within 40 days of cultivation under permissive conditions. Selection pressure is initiated after co-cultivation, and maintained until the transfer to rooting medium.

Very few variations of this general scheme have succeeded in improving transformation yields or reducing the time frame; the three-step embryogenesis process is still time-consuming, and requires repeated manipulation of explants and change in solution, thus increasing contamination risks. *Agrobacterium* transformations require a passage through in vitro regeneration of few transformed cells from an organized tissue, and thus use of selection pressure and a screenable marker is inherent to the method. Thus, in spite of its efficiency in producing stable transgenics, *Agrobacterium*-mediated DNA transfer is not considered the ideal solution for molecular farming in open-field conditions, or for the production of commercial varieties.

Table 2. Compatibility of alfalfa genotypes 1.5, 8.8 and 11.9 with *Agrobacterium tumefaciens* strains and transformation vectors for the production of transgenic embryos.

Genotype	Strain/vector	No. calluses obtained	Transformation frequency	No. embryogenic calluses	No. embryo obtained
1.5/	C58/pGA482	31	0,65	0	0
	C58/pGA4643	15	0,31	0	0
	C58/pBibKan	16	0,33	0	0
	A281/pGA482	5	0,10	0	0
	A281/pGA464	24	0,50	0	0
	A281/pBibKan	41	0,85	0	0
	LBA4404/pGA482	29	0,60	1	1
	LBA4404/pGA4643	17	0,35	0	0
	LBA4404/pBibKan	24	0,50	0	0
8.8/	C58/pGA482	16	0,34	0	0
	C58/pGA4643	22	0,46	6	41
	C58/pBibKan	0	-	0	0
	A281/pGA482	11	0,23	1	8
	A281/pGA464	21	0,44	2	4
	A281/pBibKan	17	0,35	3	5
	LBA4404/pGA482	3	0,06	1	2
	LBA4404/pGA4643	2	0,04	0	0
	LBA4404/pBibKan	4	0,08	0	0
11.9/	C58/pGA482	34	0,70	1	0
	C58/pGA4643	31	0,65	7	17
	C58/pBibKan	18	0,38	0	0
	A281/pGA482	23	0,48	0	0
	A281/pGA464	24	0,50	0	0
	A281/pBibKan	44	0,92	4	8
	LBA4404/pGA482	32	0,67	10	42
	LBA4404/pGA4643	31	0,65	4	16
	LBA4404/pBibKan	29	0,60	3	11

DIRECT DNA TRANSFER

Particle bombardment

The development of particle bombardment for the transformation of plants was first stimulated by the need to transform *Agrobacterium*-insensitive species. However, even if alfalfa is sensitive to *Agrobacterium* infection, the genotypic variability of the response, as well as the intellectual property context surrounding the use of this later transfection technique motivated the development of direct DNA transfer methods for alfalfa transfection. The particle bombardment method consists in the integration of DNA in plant cells using the acceleration of DNA-coated particles (usually gold or tungsten) into these cells. In one attempt, stable transformation could be obtained from bombardment of calli from either petiole and stem section (Pereira and Erickson, 1992). Later, another attempt of particle bombardment on suspension-cultured cells showed that bombardment was only efficient in integrating DNA transiently into cells (Brown *et al.*, 1994). It proved impossible to reproductively obtain regeneration of transgenic material. Finally, bombardment was also performed on excised immature embryos, which in proper media gave rise to polyembryogenesis (Buising and Tomes, 1995). The method requires selection pressure and skilled in vitro handling of fragile organs. However, it was shown that transgenic plants could be obtained from most commercial varieties at satisfying rates. Finally, bombardment was used to produce transgenic plants from pollen grains. The method was described as highly efficient, but produced plants which eventually lost their transgenic trait, suggesting that stable integration was not obtained, and epigenomic foreign DNA was gradually lost through repeated cell division and sexual crossing (Ramaiah and Skinner, 1997).

The method possesses the important advantage being permissive regarding the type of tissue to transform. Most organized tissues can be bombarded even if the cells possess a hard cell wall. However, the capacity of most alfalfa tissue to regenerate plants is limited, and one should assure that the tissue to be bombarded is suited to the technique, and even more, if it is appropriate for in vitro embryogenesis.

Microinjection

Microinjection of naked DNA into protoplasts, in conjunction with protoplast culture and somatic embryogenesis proved successful to produce transgenic plants from one highly regenerative genotype (Reich *et al.*, 1986). Although the method seemed tedious, relatively inefficient and extremely genotype specific at the time it was published, the development of micromanipulation methods and the isolation of new highly regenerative genotypes from commercial lines could now contribute to the improvement of such approaches. As it was described, the method required selection pressure in vitro, highly skilled handling of in vitro plant material and an efficient regeneration protocol. Recent developments in the regeneration of protoplasts for various recalcitrant species could bring alternatives to the method, especially for the isolation of transformation events, which could end up in changes in selection methods.

Electroporation

There are no reports of the use of electroporation to produce stable transgenic material in alfalfa. Unpublished results from either our laboratory or other laboratories demonstrate that electroporation is harmful to most alfalfa cellular material on which it has been tried, proving inefficient to perform DNA transfer without irreversibly harming the plant material. In one attempt, Saunders and co-workers demonstrated that electroporation could be used efficiently to introduce DNA into pollen grains, but no transgenic material was recovered from fertilization with these (J.A. Saunders, personal communication). Our laboratory has since been successful in transferring DNA to pollen grains, the main challenge being to find proper balance between germination stage and strength/duration of electrical pulse. Alfalfa pollen proved to be more sensitive to electroporation than tobacco pollen. Efficient fertilization of recipient ovules was recently achieved with electroporated pollen and yielded GUS positive F1 plants at reasonable rates (Fig. 3). GUS positive plants are currently being tested for stability of integration by PCR and Southern blotting.

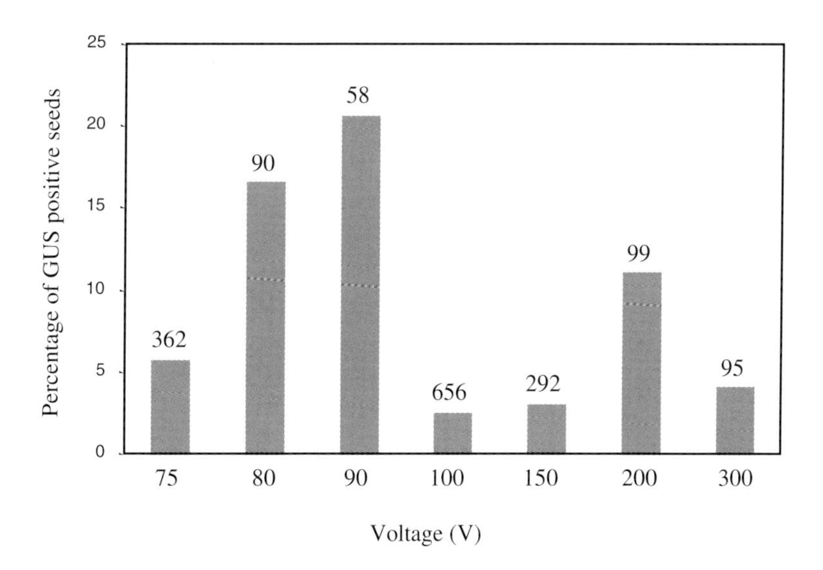

Figure 3. *GUS positive alfalfa seeds produced from electroporated pollen. The number of seeds analyzed is indicated over the columns.*

Although pollen electroporation has the disadvantage of creating transgenic egg nuclei at far greater rates than transgenic sperm nuclei, it is still an interesting method to consider for molecular farming. It does not require selection pressure, provides transgenic material within a shorter time frame than any of the other published methods, and does not require in vitro manipulation. Key issues are the identification of suitable donors, establishment of reproducible conditions for pollen production and maintenance of pollen recipients, selection of efficient fertilization pairs, establishment of pre-germination conditions and the ability to perform high throughput selection of plant material.

Other methods

There is a growing interest in alternative DNA transfection methods, particularly in the development of DNA-complexing systems that favor the endocytosis of the DNA, and target the foreign DNA to the nucleus. A variety of chemicals are now being tested in conjunction with an advanced micromanipulation method, in order to develop more efficient transformation techniques, with the ultimate goal of restricting the carriage of undesirable DNA to the minimum, while producing biosafe and stable transgenics in a short time frame.

Development of suitable direct DNA transfer techniques for plants is still at its early stages; enormous progress has been made in recent years for applications of these techniques to recalcitrant mammalian systems. Transformation of maize protoplasts using Polybrene or Lipofectin, was published by Antonelli and Stadler in 1990. More recently, it was shown that poly-L-ornithine, another polycation can act efficiently as a DNA carrier for rice protoplasts (Tsugawa *et al.* 1998). In both cases, stable transgenic plants were recovered. Recently developed transformation markers have made the use of direct DNA transfer techniques more attractive for the transformation of alfalfa, and it can be expected that stable transformation of an increasing number of plants, including alfalfa, will be performed within the years to come using DNA-complexing polymers or lipids.

POPULATION RAMP UP

Alfalfa offers access to a variety of propagation methods, each of them satisfying specific needs in term of production scale. This section presents three propagation methods that can be used for the production of grams to tons of recombinant protein. The section also presents protein manufacturing capacity of either greenhouse or field settings for alfalfa growth.

Stem propagation

The ultimate dream of the molecular farmer is to obtain access to a clonal population of any size, within a minimal time frame. This requires the ability to perform clonal propagation at large-scale. While seed production is the inevitable means of propagation for homozygous annual transgenic plants (for not having access to true clonal propagation options), faster and safer techniques can be used with perennials. Many perennial plants such as alfalfa, develop organs in which cells retain totipotency. Many herbaceous and woody perennials can thus be propagated easily by stem cuttings, provided the cutting harbors meristematic structures. Upon removal from the plant, the meristematic structure is relieved from apical dominance, and enters a fast developing stage during which both shoot and root tissue will be synthesized. Alfalfa can be stem propagated without addition of hormones, as long as the cuttings are taken from the upper part of the stem, and they are maintained in a humid environment.

A mature alfalfa stem will provide about 5 cuttings that will restore growth 3 to 4 weeks after removal from the stem. These cuttings can in turn provide 5 cuttings 4 weeks after transplantation. For large-scale propagation, cuttings can be rooted and allowed to develop at a density of more than 1000 stems per square meter. A mature alfalfa stand will bear *ca* 600 to 800 stems per square meter, originating from 100-150 individual plants. A greenhouse alfalfa stand can be grown under higher densities, up to 1000 stems per square meter; thus one square meter of alfalfa cuttings provides for 6-10 square meters of an alfalfa production stand.

Features of alfalfa population ramp up using stem propagation are presented in Table 3. This type of propagation has its limits, but since a 110 square meter greenhouse will hold 11000 plants – that is 110000 stems – and provide 77 kg of dry matter per harvest at 0,5 g recombinant protein per 100 g total SDS extractable protein (24% of dry matter), it is expected that this greenhouse could sustain the production of about 450 g of purified recombinant protein annually. Starting from a single plant, the time required to produce this clonal population is 4 months.

Stem propagation will therefore be the method of choice for production schemes within the boundaries of 1-5 kg annual demand for recombinant product, under greenhouse conditions, or under controlled irrigated field conditions. Stem propagation can be automated to a certain extent; it has been done for in vitro propagation particularly in eastern countries, and similar automation systems are being evaluated for their potential value for large-scale stem propagation. Hand propagation is time consuming, but still a very efficient means of producing clonal populations from single transgenics. It is also a strategy that is not available for most annual crops, including maize, soybean, peas and wheat.

Stem propagation also provides an easy and cost-effective means of increasing temporarily a population to a size suitable for the production of gram levels of recombinant protein; this is a key asset when considering the time required for product testing and homologation. Using transgenic alfalfa for the production of a therapeutic implies that a small clonal population can be produced with minimal expenses, to support product demand, through all phases of testing and homologation, with the guarantee that a production-scale identical population can be produced with minimal delays, to meet marketing demands.

Table 3. *Methods and conditions for increasing the population of alfalfa plants according to production scales of recombinant protein (assuming an expression level of 0,5% of total soluble proteins and purification yields going from 10% (10 mg level) to 60% (10 g level), and 90% (2 kg-1 ton level).*

First harvest* (g molecule)	Interval to First delivery (months)	Nb of Plants / area	Yield/harvest (5-6 wks) (g molecule)	Propagation method	Growing premises
10 mg	4	70 / 1m^2	50 mg	Stem cutting	Greenhouse
100 mg	6	900 9 m^2	1g	Stem cutting	Greenhouse
10 g	8	11 000 / 110 m^2	45 g	Stem cutting	Greenhouse
150 g	10	140 000 / 0.14 ha	600 g	Stem cutting	Greenhouse
700 g	13	1 000 000 / 1 ha	6,3 kg	Somatic embryogenesis	Greenhouse
100 kg	30	36 ha	200-400 kg/season	Seeds	Field
1t	30	360 ha	2-4 t/ season	Seeds	Field

* First harvest from immature plants, in the greenhouse two growth cycles are needed to get mature plants which will yield greater biomass every 5 to 6 weeks.

Somatic embryogenesis

The endogenous ability to produce embryos from somatic tissue has been exploited for the production of clonal material at large-scale through a process described as artificial seed production. It was initially developed to avoid the complex heredity of alfalfa with the objective of fixing agronomic traits in true hybrid varieties. To this day, the use of the technology has not been extensive in alfalfa improvement. However, elements of this technology are now being used to generate clonal populations at large-scale from initial transgenic lines for molecular farming.

Typically, petioles from transgenic plants are sterilized and placed in liquid cultures; through a period of 2-3 weeks, cells will be liberated from totipotent dividing cell layers within the vascular tissues of the petiole. This cell suspension is then layered onto a solid culture medium on which embryos will develop during the 3 following weeks. In ideal conditions, more than 50 embryos will be produced from one ml of cell suspension; 3000-5000 embryos can therefore be produced from a 50 ml culture vessel, and when conditions for initiating liquid cultures are followed carefully, this system can be scaled-up easily to produce more than one million embryos from three 4-liter flasks. These embryos can be dehydrated following an ABA treatment and frozen for long-term storage. Viability issues of the dehydrated embryos, and the potential of similar methods for large-scale implantation of clonal alfalfa stands have been discussed in earlier reports (Senaratna *et al.*, 1989; McKersie *et al.*, 1989; Senaratna *et al.*, 1990).

In ideal establishment conditions (greenhouse operations), one million plants, that is, the plants regenerated from embryos produced in three 4-liter flasks, suffices to cover a one-hectare production plot. Combined with the perenniality of the crop, this one-million-embryo lot provides a production lot that will survive 8-9 harvests per year for over ten years. Embryos can be grown as for seeds in micro cavities and transplanted to the field, or sown directly in humid soil. Although embryogenesis from somatic tissue requires installations and handling which is not typical of conventional propagation for a forage crop, it is the method of choice for producing a clonal population at large-scale. Simple laboratory facilities will allow production of the order of one million embryos, which can be dehydrated and stored with minimum care. Open field planting or sowing efficiency depends on the viability of the embryos and soil condition, type and humidity. Considering a germination rate of 50%, the production of one million plants could be obtained within 13 months, and this population would produce enough biomass to yield 6.3 kg of recombinant protein per harvest (Table 3). This period includes 8 weeks for growing a first set of plants from which the petioles needed for the production of the cell suspension cultures will originate.

Seed Production

While stem propagation and embryogenesis are ideal to produce clonal populations, they require skilled manipulations, which bring limitations to applications at ultra large scale. Seed production is still currently the only method available for the production of alfalfa populations that could supply tons of recombinant proteins per year.

Since alfalfa is very sensitive to inbreeding and has limited self-fertility, seed production of transgenic plants is most easily accomplished after one or more backcrosses to an elite, genetically broadbased population that is ideally adapted to the region targeted for field production of said transgenic plants. After each backcross it will be necessary to select for the transgenic genotype or phenotype. It is possible to produce two sexual generations of alfalfa per year in the greenhouse. The transgenic segregants from such a backcrossing program will be simplex (Axxx) for the transgene.

These simplex individuals can be intercrossed in the field to produce larger quantities of seed. Proper geographic isolation will be required to prevent gene escape and control purity of the transgenic trait. Reliable alfalfa seed production in the field is limited to areas with a warm dry summer. Alfalfa seed production is generally a specialized farming practice and requires intensive management of pollinators, irrigation, and insect pests. Cool weather during pollination or rain during seed maturation/harvest can severely limit seed yield or seed quality. Commercial alfalfa seed production in North America is concentrated in California and the Pacific Northwest in the U.S. and in the western Prairie provinces in Canada. In the Pacific Northwest and California it is possible to produce seed in the year of stand establishment. For example, seed production in Idaho of transgenic alfalfa plants transplanted to the field in April at a density of 5 plants/m^2 would likely yield an average of about 20 grams of seed per plant when harvested in September. Thus an intercross of 1000 transgenic alfalfa plants in the field could yield approximately 20 kg of seed in the establishment year.

An intercross of simplex individuals in an autotetraploid will yield a population in which 75% of the progeny express the transgenic protein. Breeding systems using two independent transgenic events have been devised that yield populations with >90% of the progeny expressing the transgenic protein (McCaslin, unpublished).

Additional generations of seed production could be used to further increase the amount of seed of transgenic populations. Alfalfa seed production fields can be successfully established with a seeding rate of 0.5 – 1 kg/ha. In the establishment year, seed yield in the Pacific Northwest or California averages about 500 kg/ha. Seed yield in the second and third year averages about 700 kg/ha.

Altogether, using seed production, we calculate that, after 30 months, 2 to 4 tons of recombinant proteins might be purified each year from the foliage produced by 360 hectares of alfalfa (Table 3).

GROWTH, HARVEST AND INITIAL PROCESSING

Since alfalfa will regrow after consecutive harvests, and will usually form a larger number of stems from basal buds than that of the previous growth, final density of the stand is quite independent of initial sowing or planting density.

Following planting of rooted cuttings, production stands are obtained after two initial cuts, which are performed to increase growth vigor and density of the canopy. If seeding density is high (over 14 kg per hectare) and performed late summer, production can be initiated the following year of growth. Depending on genotype, stand density, fertilization and other growth conditions, regrowth can proceed over 4 to 5 weeks before flower primordia appear. At this stage, the average stem weight is ca 3.5 g, and thus harvest will yield 3.5 kg fresh matter per square meter in the greenhouse, and 7-10 tons fresh matter per hectare in the field. Under greenhouse production, 8-10 harvests can be performed within a year, thus yielding ca 28-35 kg per square meter. Depending on the geographical location of the production plots, 3-5 harvests can be performed per year in the field, providing ca 8-12 tons of dry matter per year.

Greenhouse settings are suitable for continuous production projects for which demand of the recombinant protein is less than 10 kg a year. Under these cultivation conditions, harvest can be performed non-mechanically with minimal apparatus. For larger exploitation, field harvest is performed with a standard forage harvester, with modifications to the blade system brought about to minimize stem breakage and shearing until further processing. Specially adapted equipment can cut and collect tons of fresh matter per hour on a continuous basis. This machinery has been used extensively for ultra-large scale exploitation of non-transgenic alfalfa in France and the USA. In minimizing stem and leaf damage through the process of harvesting, proteolysis and oxidation are kept to a minimum. Freshly cut foliage and stems are transported to processing facilities.

The capacity to produce at large- (hectare) and ultra-large scale (1000 hectares and more) is key to the development and acceptance of molecular farming. One advantage of the alfalfa system, which for long was seen as a disadvantage, is that the protein is produced in a tissue, which is totally removed from the field upon harvest. Because of the complexity of the leaf endogenous protein components and because of the complex metabolism of leaves the constraints of proteolysis and oxidation is a reality to this production system. One other reality is the fact that processed products of the raw green matter from alfalfa have been marketed for years, and are commonly used as feed products, thus offering both a known industrial process portfolio for the extraction of juices and handling of solid fractions, and a well described and accepted marketing niche for extraction co-products. Leaf protein content is a lot more varied and balanced than that of grains or other seeds; this complexity provides the basis for diversity of applications; alfalfa is easily seen as a whole feed crop, rather than a limited specialized staple or additive. Following is a general outline of the industrial processing of alfalfa foliage using the wet fractionation method, which is currently adapted for recovery of valuable recombinant products from transgenic foliage. Main steps and parameters of the wet fractionation method are also summarized in Fig. 4.

INDUSTRIAL PROCESSING

Extrusion of the green juice

Soon after harvest, alfalfa is chopped and pressed through a compacting screw for separation of two main fractions equal in size, a highly fibrous fraction containing *ca* 15% protein and a liquid colloidal fraction containing the other cellular components. Soluble and insoluble elements are found in this latter fraction and from a nutritional perspective, it contains mostly proteins, vitamins and pigments. The co-product of this green juice, the lignocelluloses components from leaves and stems, are dried, pelleted, and commercialized for animal nutrition. Its nutritive value can be improved by fragmentation of the cellulosic parts of the stems to increase their digestibility.

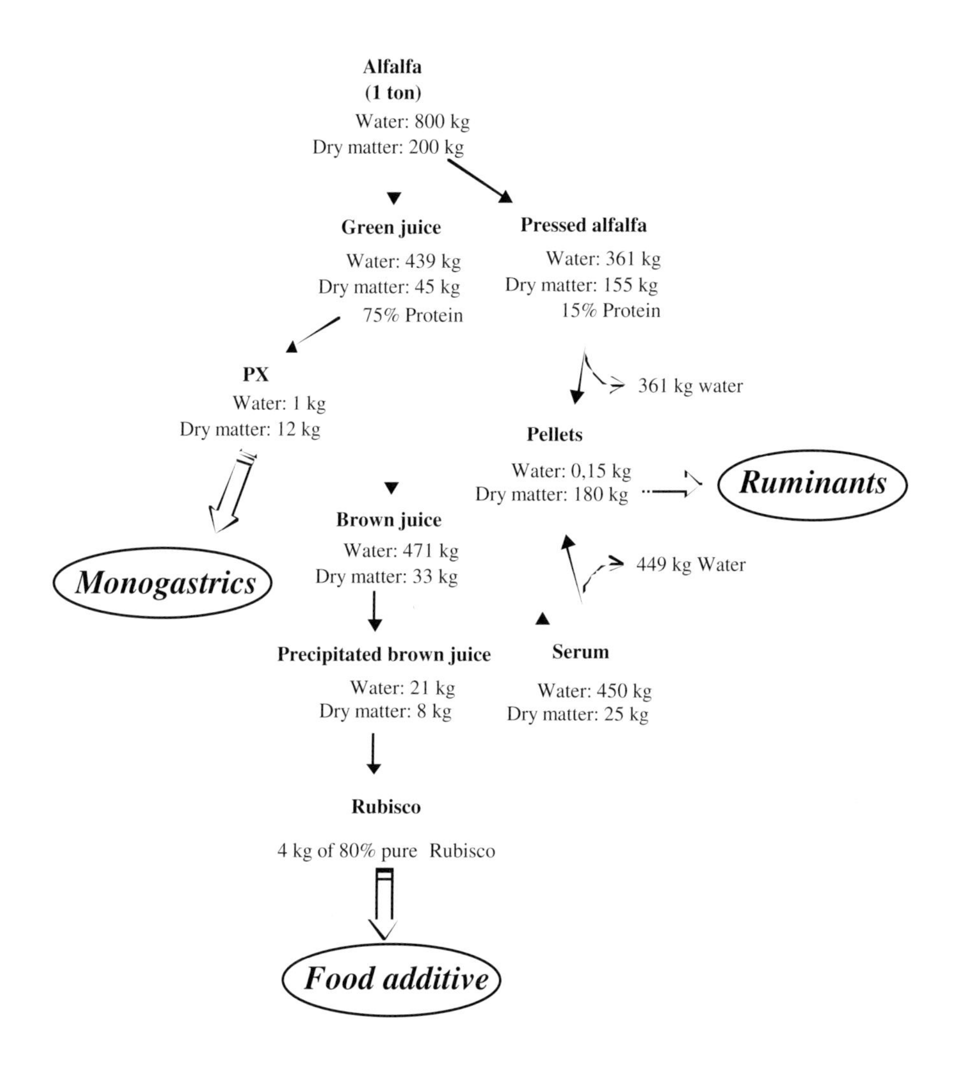

Figure 4. *Steps of alfalfa processing toward the purification of Rubisco.*

Heat coagulation

The green juice produced by extrusion, is adjusted to pH 8.5 to slow down the action of phenoloxydases, preheated, and brought to 57°C by steam injection. This heating step provokes coagulation of most proteins, pigments, liposoluble vitamins, lipids and some mineral salts. Adaptations under work for this process have demonstrated that efficient coagulation can be performed at much lower temperatures (47°C), thus preserving both a higher structure and activity of both coagulated and residual proteins; it also protects its pigment content from degradation.

Sedimentation & clarification

The coagulum (green humid paste containing the main nutriments) is separated from the rest of the solution (brown serum) by centrifugation and then dried on a "fluidized bed". A further pelleting step allows conditioning of the product for its consumption by animals such as hens, chickens, calves or piglets. This product is commercialized under the name of PX, which stands for "proteins and xanthophylls".

Extraction and purification of Rubisco

The brown juice still contains most soluble proteins, from which Rubisco is by far the most abundant. A Rubisco-rich fraction is currently being extracted, partially purified and conditioned from the brown juice through a proprietary process and is currently marketed as a food additive while other more refined applications and commercial niches are being developed. Amino acid composition, solubility and general physico-chemical properties of these extracts or of their hydrolysates make them ideal for components of fermentation or cell culture media. Purified Rubisco is currently produced at more than one ton per day through current processes.

Applications for molecular farming

Scale, reliability and efficiency of the wet fractionation process make it a perfect scale-up strategy for molecular farming with alfalfa. Our improvements of the current industrial method have already shown that the whole process can be easily adapted for the production and extraction of IgG, which are soluble proteins; to this date, small-scale trials have demonstrated that mAb C5-1 is stable through all processes involved in the preparation of Rubisco and co-products, as described above, if coagulation temperature is kept below 52°C at all times. The process of producing IgGs with transgenic alfalfa thus produces purified IgGs, together with a Rubisco-enriched fraction, a protein/pigment-enriched fraction, and a fibrous fraction at 15% protein, all products being suitable for market. A case study describing the issues involved in IgG production in alfalfa is described below.

The combined quality of alfalfa as animal feed and the versatility of the recovery processes developed for alfalfa (see below) have broadened the range of its applications far beyond the exemplified factory for soluble biologicals. Following are examples of applications under development.

Oral delivery for ruminants

Alfalfa is the basic fodder feed for ruminants all over the world. Although its protein content and quality usually calls for complementation with forage grasses and high-energy grains, it is widely used as basic diet for milking cows, livestock and sheep, as dried bailed forage or as silage. Ruminants, like most other animals kept under intensive production suffer from a variety of minor illnesses, most of which are related to the gastro-intestinal tract, and impair growth and normal development. This is true at all developmental stages, although more significant at the early stages. Some of the GI infections or inflammatory conditions have been controlled by proper use of antibiotics, which reduce the proliferation of the causal agent. However these treatments are often inefficient, because the animals are under boosting diets, and chronic inflammation persists even when the initial causal agent has disappeared; this condition both restricts normal GI functions and leaves room for opportunistic infections. There is a significant loss each year in the meat industry alone, due to improper weight gain as a consequence of chronic GI inflammation.

We are now involved in the development of different prophylactic and therapeutic strategies based on active elements that could be produced by transgenic alfalfa, and remain active through ingestion of the transgenic foliage. Target molecules for these strategies range from neutralizing antibodies for microbes and cytokines, to ligand proteins for various toxins.

N-glycosylation of alfalfa proteins

The formation process, the structure and the function of naturally produced plant and mammalian N-linked glycans share similarities. For example, in both organisms, the glycan facilitates the correct folding of the glycoprotein and protects the protein against proteolytic degradation. However, some differences in the fine structure of plant and mammalian N-glycans have important consequences on the capacity to produce mammalian proteins in alfalfa. Plant glycosylation differs from mammalian glycosylation by the presence of an α(1,3)-fucose residue instead of the α(1,6)-fucose on the proximal GlcNAc, and a ß(1,2)-xylose linked to the ß-mannose. Hence, for some mammalian proteins for which the structure of the glycan is a determinant of the activity of the glycoprotein, it appears necessary to determine if the alfalfa produced protein possesses similar activity to its naturally occurring sister, and that for each individual recombinant protein that is produced.

A second issue concerns the allergenicity of plant-produced glycoproteins. Some observations suggest that the N-glycan present on an allergenic protein is involved in the reactivity of the glycoprotein with specific IgE antibodies. For this reason, the difference in N-glycan structure between the alfalfa-produced and the original protein has to be investigated to determine its potential allegenicity. Hopefully, the experience acquired in the near future will help us determine the specific residues involved in the reactivity and find tools to overcome this potential problem.

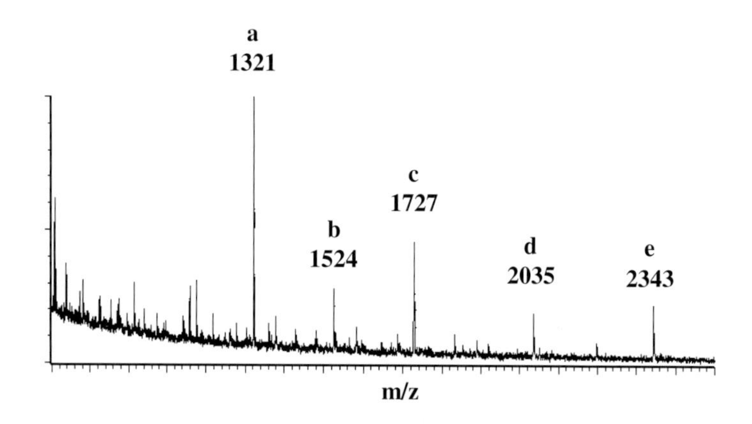

Figure 5. *MALDI-TOF mass spectrum of N-glycans released from alfalfa proteins by PNGase A. The spectrum was recorded after addition of CsCl to the sample to obtain single $M+Cs^+$ ion for each oligosaccharide. $M+Cs^+$ ions were then assigned to N-glycans* ***a*** *to* ***e*** *represented in Fig. 6.*

As in other eukaryotes, the N-glycosylation of plant proteins starts by the transfer in the endoplasmic reticulum (ER) of an oligosaccharide precursor on specific Asn residues. Then, the processing of plant N-linked glycans occurs along the secretory pathway as the glycoprotein moves from the ER through the Golgi apparatus to its final destination. Glycosidases and glycosyltransferases located in the ER and in the Golgi apparatus successively modify the oligosaccharide precursor to high-mannose-type N-glycans and then into complex-type N-glycans. Plant complex-type N-glycans are characterized by the presence on the $Man_3GlcNAc_2$ core of ß(1, 2)-xylose residue linked to the ß-mannose and/or an α(1, 3)-fucose residue linked to the proximal glucosamine residue, and/or terminal GlcNAc residues or Galß1-3)[Fucα1-4)]GlcNAc oligosaccharide sequences, named $Lewis^a$ (Fitchette-Lainé *et al.*, 1997, Melo *et al.*, 1997).

We have recently analyzed the N-glycosylation pattern of alfalfa proteins. A preliminary analysis by immunodetection on blots, performed on a crude protein extract, has shown that alfalfa glycoproteins are immunodetected with antibodies specific for glycan epitopes that are found in complex-type plant N-glycans, i.e. antibodies specific for the α(1, 3)-fucose residue, antibodies specific for the ß(1, 2)-xylose residue (Faye *et al.*, 1993) or antibodies specific for $Lewis^a$ antigens (Fitchette-Lainé *et al.*, 1997).

A detailed structural characterization of these N-linked glycans was then carried out by mass spectrometry analysis. In this respect, a pool of N-glycans was released from an alfalfa protein extract using PNGase A as previously described (Fitchette *et al.*, 1999; Bardor *et al.*, 1999) and analyzed by Matrix-Assisted Laser Desorption Ionisation-Time of Flight mass spectrometry (MALDI-TOF MS). Fig. 5 shows the MALDI-TOF MS of the pool of N-glycans isolated from alfalfa plants. On the basis of their molecular weight, four major $M+Cs^+$ ions were detected and assigned to structures **a** to **e** represented in Fig. 6. The structure of these N-glycans was confirmed by enzyme sequencing combined with MALDI-TOF mass spectrometry as previously reported (Bakker *et al.*, 2000). Minor complex oligosaccharides lacking one fucose residue and high-mannose-type N-glycans ranging from $Man_5GlcNAc_2$ to $Man_9GlcNAc_2$ were also detected in the pool of N-linked glycans. As a consequence, both on-blot and mass spectrometry analysis of the N-glycans show that proteins from alfalfa plants are N-glycosylated mostly by matured oligosaccharides having core α(1, 3)-fucose and core ß(1, 2)-xylose residues as well as $Lewis^a$ antigens as reported for tobacco plants (Fitchette *et al.*, 1999; Bakker *et al.*, 2000).

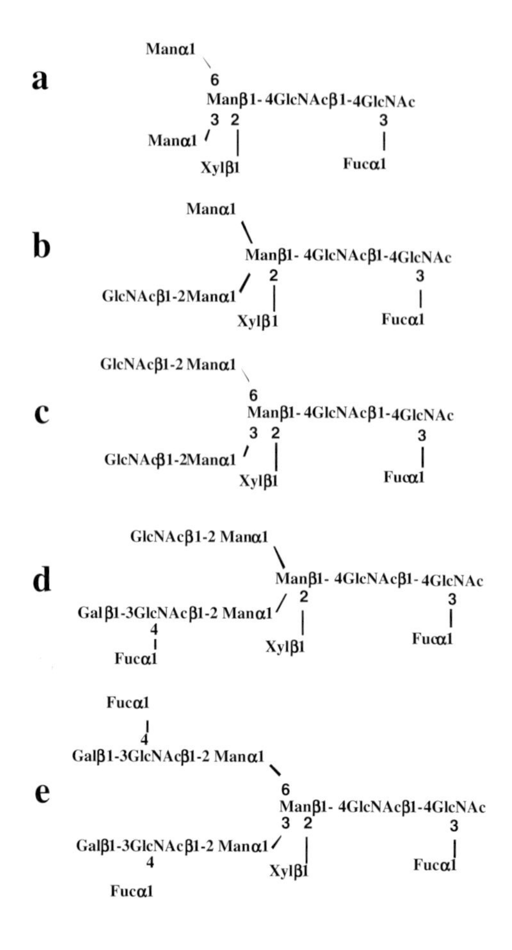

Figure 6. *Structure of N-glycans isolated from alfalfa plants.*

CASE STUDIES

The two studies presented here illustrate the diversity of recombinant protein that can be produced in alfalfa. First, the production of an anti-human immunoglobulin (IgG) is presented. IgGs are complex glycosylated heterotetrameric proteins composed of two different subunits. Correctly assembled mature antibodies contain two light chains of 214 a.a. and two heavy chains of 446 a.a. held together by disulfide bonds. Also presented is the production of Interleukin-2 (IL-2), a 153 a.a. long peptide containing 20 a.a. of signal peptide which targets the protein to the extra-cellular space. Mature IL-2 is a single chain polypeptide of 15,000 to 18,000 Daltons depending on the degree of glycosylation, and the only disulfide bond present on mature IL-2 is essential for its biological activity.

Human IgG production using alfalfa

An increasing number of monoclonal antibodies (mAb) are currently being developed for therapeutic and diagnostic use. The hybridoma technology currently used for the production of mAbs has a limited scale-up capacity, is costly, and production is unstable in long-term cultures. C5-1 described herein is an anti-human IgG used as a key and major component in an agglutination reagent for blood cell phenotyping, and cross matching between donors and receivers.

Transgenic alfalfa plants were produced that contained either the cDNA encoding the light or the heavy chain of the antibody. Plants accumulating both the light and the heavy chain of the antibody were obtained by intercrossing these parental mono- transgenic plants. Fig. 7 presents a Western blot analysis of the progeny revealing the presence of both the light and the heavy chain in F1 plants. The plants harboring both transgenes contained fully assembled heterotetrameric C5-1 as well as intermediates of assembly (HL, H2, H2L, and H2L2).

Specific activity of the alfalfa derived C5-1 was compared to that of the original antibody; C5-1 mAb was first purified from alfalfa leaf extracts using expanded bed affinity chromatography (STREAMLINE rProteinA). A single C5-1 peptide was obtained, and the activity of purified C5-1 was measured using a specific ELISA and antigen-binding assay equivalent to that of hybridoma-derived C5-1 (Table 4). Stability of the plant C5-1 was also monitored in planta and in the blood stream of mice. It was shown that the antibody was stable in the drying hay (Fig. 8) after a 12-week storage at room temperature. We have compared the degradation rates of the plant-derived C5-1 with that of the hybridoma-derived antibody. As shown in Fig. 9, hybridoma- and plant-derived C5-1 antibodies had a half-life of 3 days in the blood stream of mice.

Production of C5-1 anti-human IgG was the first demonstration of the capacity of a perennial plant to produce a complex multimeric recombinant protein. The study demonstrates that the efficiency of the plant derived C5-1 is equivalent to that of its hybridoma-derived counterpart. It also shows that the plant bioreactor can be dried without affect on the stability of C5-1, making the foliage production system equivalent to a seed production system with regards to storage and transportation costs.

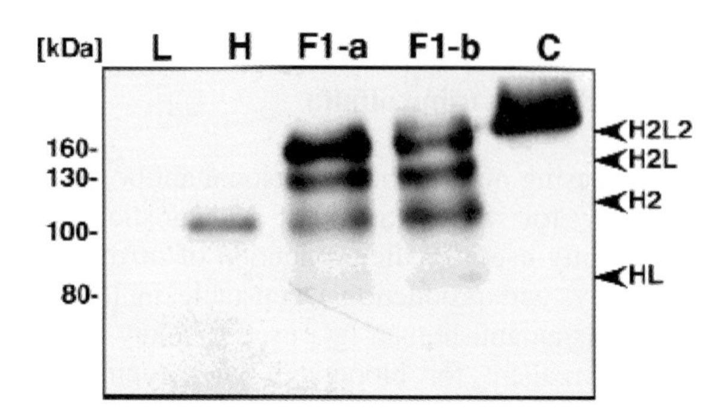

Figure 7. *Protein blot analysis of C5-1 light and heavy chain assembly in alfalfa. Total protein extracts from plants producing the light chain (L), the heavy chain (H), and both chains (F1-a and F1-b) were analyzed and compared to purified C5-1 from a hybridoma. From Khoudi et al. (1999) Biotechnology and Bioengineering. © 1999 John Wiley & Sons, Inc. Reprinted by permission of Wiley-Liss, Inc., a subsidiary of John Wiley & Sons, Inc.*

Table 4. *Comparison of hybridoma- and alfalfa-derived C5-1 characteristics*

Extract	Specific activity (OD/100ng)	True affinity (K_{DS})
C5-1 from hybridoma	0.235 $\pm$ 0.020	4.6 X 10^{-10} M
C5-1 from alfalfa	0.267 $\pm$ 0.080	4.7 X 10^{-10} M

From Khoudi et al. (1999) Biotechnology and Bioengineering. © 1999 John Wiley & Sons, Inc. Reprinted by permission of Wiley-Liss, Inc., a subsidiary of John Wiley & Sons, Inc.

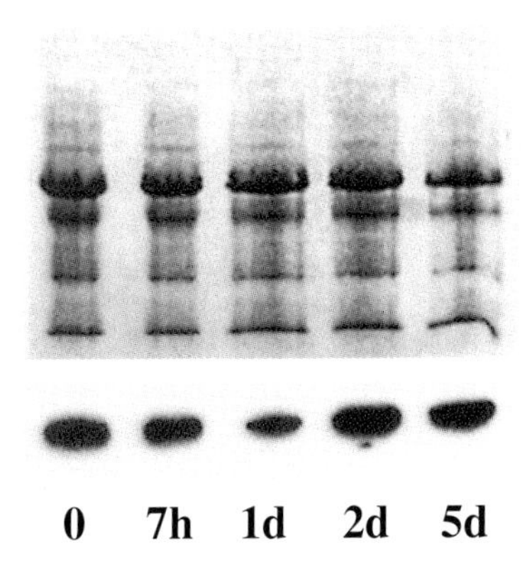

Figure 8. *Stability of C5-1 in alfalfa. Transgenic alfalfa extracts were left up to 5 days at room temperature, total soluble proteins were extracted (upper part) and analyzed for the stability of C5-1 by protein gel blot (lower part). From Khoudi et al. (1999) Biotechnology and Bioengineering. © 1999 John Wiley & Sons, Inc. Reprinted by permission of Wiley-Liss, Inc., a subsidiary of John Wiley & Sons, Inc.*

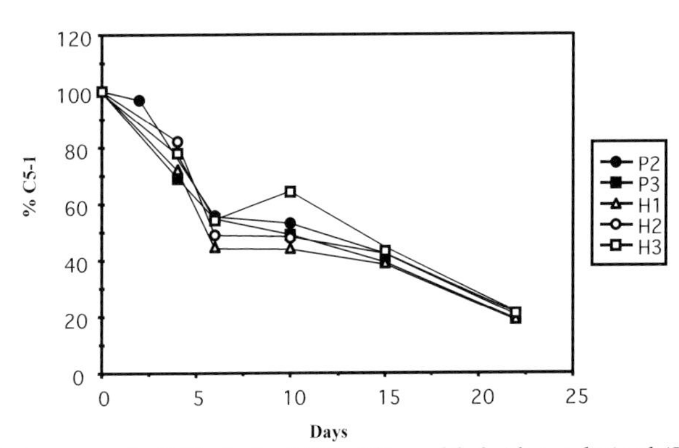

Figure 9. *Degradation rate of alfalfa derived (P2, P3), and hybridoma derived (H1, H2, H3) C5-1 in the blood stream of mice. From Khoudi et al. (1999) Biotechnology and Bioengineering. © 1999 John Wiley & Sons, Inc. Reprinted by permission of Wiley-Liss, Inc., a subsidiary of John Wiley & Sons, Inc.*

Human Interleukin-2 production using alfalfa

Interleukin-2 (IL-2) is naturally produced and released by T-helper cells of the immune system in reaction to infection. IL-2 induces the maturation and division of T-helper cells and promotes their fighting ability against invading organisms. IL-2 finds its therapeutic use in the immune-based therapy against cancer, and is believed to increase life quality and expectancy of HIV infected people.

A PCR-amplified human IL-2 coding region, including its own signal peptide was linked to a 35S promoter in a transcriptional fusion. Transgenic alfalfa showed accumulation of mature IL-2 in leaf tissue. The analysis of IL-2 accumulation level indicated that low amounts of IL-2 were present, typically at a maximum level of about 0.001 % of total soluble proteins (TSP). This level of expression was enough to stimulate the growth of cultured IL-2 dependent cells (CTLL-2). Table 5 shows the biological activity of IL-2-alfalfa extracts on CTLL-2 cell growth.

Table 5. *Cell growth promoting activity of IL-2-containing extracts.*

Protein extract	Il-2 accumulation level (% soluble proteins)	Biological activity (U / mg protein)
Crude 35S-IL2-2.c	0,0008	$1,5 \times 10^5$
Crude 35SIL2-2.d	>0,0014	4×10^5
Crude 35S-IL2-2.e	0,0012	$4,8 \times 10^4$
Human IL-2 (control)	-------	$1,3 \times 10^7$

This study showed the ability of the perennial alfalfa plant to produce a highly active signaling protein of the human immune system. Although the accumulation level was relatively low (we cannot exclude the possibility that the protein is toxic for the plant at higher concentration), the crude plant extracts showed a substantial biological activity on IL-2 dependant cells. It is not known at present if the signal peptide was correctly recognized and processed by the plant secretion pathway. Further attempts to target accumulation to other cell compartments resulted in similar accumulation levels, except for the endoplasmic reticulum where clear signs of toxicity were monitored.

REFERENCES

Antonelli NM and Stadler J. (1990) Genomic DNA can be used with cationic methods for highly efficient transformation of maize protoplasts. Theor. Appl. Genet. 80: 395-401.

Atanassov A and Brown DCW. (1984) Plant regeneration from suspension culture and mesophyll protoplasts of *Medicago sativa* L. Plant Cell Tiss. Org. Cult. 3: 149-162.

Bagga S, Sutton D, Kemp JD, Sengupta-Gopalan C. (1992) Constitutive expression of the beta-phaseolin gene in different tissues of transgenic alfalfa does not ensure phaseolin accumulation in non-seed tissue; Plant Mol. Biol. 19: 951-958.

Bakker, H, Bardor M, Molhoff, J, Gomord V, Elbers I, Stevens L, Jordi W, Lommen A, Faye L, Lerouge P, and Bosch D. Humanised glycans on antibodies produced by transgenic plants. Proc. Natl. Acad. Sci. USA, in press (2000).

Bardor M, Faye L, and Lerouge P. Analysis of the N-glycosylation of recombinant glycoproteins produced in transgenic plants. Trends Plant Sci., 9, 376-380 (1999).

Benmoussa M (1999) Production d'une glutéine à faible poids moléculaire dans les feuilles de la luzerne et les tubercules de la pomme de terre. PhD thesis. Laval University, Québec, Canada.

Bingham ET. (1978) Maximizing heterozygosity in autopolyploids p. 471-489 *In* W.H Lewis (ed.) Polyploidy: Biological Relevance. Plenum Press, New York.

Blaydes DF. (1966) Interaction of kinetin and various inhibitors in the growth of soybean tissue. Physiol. Plant. 19: 748-753.

Brown DC, and Atanassov A. (1985) Role of genetic background in somatic embryogenesis in Medicago. Plant Cell Tiss. Org. Cult. 4: 111-122.

Brown DCW, Tian L, Buckley DJ, Lefebvre M, McGrath A, Webb J. (1994) Development of a simple particle bombardment device for gene transfer into plant cells. Plant Cell Tiss. Org. Cult. 37: 47-53.

Buising CM and Tomes D. (1995) Methods of regeneration of Medicago sativa and expressing foreign DNA in same. US Pat. # 5,994,626.

Chabaud M, Passatore JE, Cannon F, Buchanan-Wollaston V. (1988) Parameters affecting the frequency of kanamycin resistant alfalfa obtained by *Agrobacterium tumefaciens* mediated transformation. Plant Cell Rep. 7: 512-516.

Chalfie M, Tu Y, Euskirchen G, Ward WW, Prasher DC. (1994) Green fluorescent protein as a marker for gene expression. Science 263: 802-805.

Davis SJ and Viestra RD. (1998) Soluble, highly fluorescent variants of green fluorescent protein (GFP) for use in higher plants. Plant Mol. Biol. 36:521-528

Deak M, Kiss GB, Koncz C, Dudits D. (1986) Transformation of Medicago by Agrobacterium mediated gene transfer. Plant Cell Rep. 5: 97-100

Desgagnés R, Laberge S, Allard G, Khoudi H, Castonguay Y, Lapointe J, Michaud R, Vézina LP. (1995) Genetic transformation of commercial breeding lines of alfalfa (*Medicago sativa*). Plant Cell, Tiss. Org. Cult. 42: 129-140.

Faye L, Gomord, V, Fitchette-Lainé, A-C, and Chrispeels MJ. Affinity purification of antibodies specific for Asn-linked glycans containing α1-3 fucose or ß1-2 xylose. Anal. Biochem., 109, 104-108 (1993).

Fitchette AC, Cabanes-Macheteau M, Marvin L, Martin B, Satiat-Jeunemaitre B, Gomord V, Crooks K, Lerouge P, Faye L, and Hawes C. Biosynthesis and immunolocalization of Lewis A-containing N-glycans in the plant cell. Plant Physiol., 121, 333-343 (1999).

Fitchette-Lainé A-C, Gomord V, Cabanes M, Saint Macary M, Foucher, B, Michalski J-C, Hawes C, Lerouge P, and Faye, L. N-glycans harboring the lewis a epitope are expressed at the surface of plant cells. Plant J., 12, 1411-1417 (1997).

Groose RW and Bingham ET. (1984) Variation in plants regenerated from tissue culture of tetraploid alfalfa (*Medicago sativa*) heterozygous for several traits. Crop Sci. 24: 655-658.

Haseloff J, Siemering KR, Prasher DC, Hodge S. (1997) Removal of a cryptic intron and subcellular localization of green fluorescent protein are required to mark transgenic Arabidopsis plants brightly. Proc. Natl. Acad. Sci. USA 94: 2122-2127.

Heim R, Prasher DC, Tsien RY. (1994) Wavelength mutations and posttranslational autoxidation of green fluorescent protein. Proc. Natl. Acad. Sci. USA 91: 12501-12504.

Hill RR. (1976) Response to inbreeding in alfalfa populations derived from single clones. Crop Sci. 16: 237-241.

Holbrook LA, Reich TJ, Iyer VN, Haffner M, Miki BL. (1985) Induction of efficient cell division in alfalfa protoplasts. Plant Cell Rep. 4:229-232.

Jefferson RA, Kavanagh TA, Bevan MW. (1987) GUS fusions: ß-glucuronidase as a sensitive and versatile gene fusion marker in higher plants. EMBO J. 6: 3901-3907.

Jobling SA and Gehrke L. (1987) Enhanced translation of chimaeric messenger RNAs containing a plant viral untranslated leader sequence. Nature 325: 622-625.

Johnson LB, Stuteville DL, Higgins RK, Skinner DZ. (1981,) Regeneration of alfalfa (*Medicago sativa* cultivar saranac) plants from protoplasts of selected Regen S clones. Plant Sci. Lett. 20: 297-304.

Kao KN and Michayluk MR. (1979) Plant regeneration from mesophyll protoplasts of alfalfa. Z Pflanzenphysiol. Bd. 96.S: 135-141.

Khoudi H, Laberge S, Ferullo JM, Bazin R, Darveau A, Castonguay Y, Allard G, Lemieux R, Vezina LP. (1999) Production of a diagnostic monoclonal antibody in perennial alfalfa plants. Biotechnol. Bioeng 64: 135-143.

Khoudi H, Vézina LP, Mercier J, Castonguay Y, Allard G, Laberge S. (1997) An alfalfa rubisco small subunit homologue shars cis-acting elements with the regulatory sequences of the RbcS-3A gene from pea. Gene 197:343-351.

Kilcher MR and Heinrichs DH. (1974) Contribution of stems and leaves to yield and nutrient level of irrigated alfalfa at different stages of development. Can. J. Plant Sci. 54: 739-742.

Matheson SL, Nowak J, Maclean NL. (1990) Selection of regenerative genotypes from highly productive cultivars of alfalfa. Euphytica 45: 105-112.

McKersie BD, Senaratna T, Bowley SR, Brown DCW, Krochko JE, Bewley JD. (1989) Application of artificial seed technology in the production of hybrid alfalfa (*Medicago sativa* L.). In Vitro Cell. Dev. Biol. 25: 1183-1188.

Melo NS, Nimtz M, Conradt H, Fevereiro PS, and Costa J. Identification of the human Lewis[a] carbohydrate motif in a secretory peroxidase from a plant cell suspension culture (*Vaccinium myrtillus* L.). FEBS Lett., 415, 186-191 (1997).

Mowat, DN, Fulkerson RS, Tossel WE and Winch JE. (1965) The *in vitro* digestibility and protein content of leaf and stem portions of forages. Can J. Plant Sci. 45: 321-331.

Murashige T and Skoog F. (1962) A revised medium for rapid growth and bio assays with tobacco tissue cultures. Physiol. Plant. 15: 473-497.

Narvàez-Vàsquez J, Orozco-Càrdenas ML, Ryan CA. (1992) Differential expression of a chimeric CaMV-tomato proteinase Inhibitor I gene in leaves of transformed nightshade, tobacco and alfalfa plants. Plant Mol. Biol. 20: 1149-1157.

Nowak J, Matheson S, Maclean NL, Havard P. (1992) Regenerative trait and cold hardiness in highly productive cultivars of alfalfa and red clover. Euphytica 59: 189-196.

Odell JT, Nagy F, Chua NH. (1985) Identification of DNA sequences required for activity of the cauliflower mosaic virus 35S promoter. Nature 313: 810-812.

Pereira LF and Erickson L. (1992) Stable transformation of alfalfa (*Medicago sativa* L.) by particle bombardment. Plant Cell Rep. 14: 290-293.

Ramaiah SM and Skinner DZ. (1997) Particle bombardment: A simple and efficient method of alfalfa (*Medicago sativa* L.) pollen transformation. Cur. Sci. (Bangalore) 73:674-682.

Reich TJ, Iyer VN, Miki BL. (1986) Efficient transformation of alfalfa protoplasts by the intranuclear microinjections of Ti plasmids. Biotechnology 4:1001-1004.

Schenk RU and Hilderbrandt AC. (1972) Medium and techniques for induction and growth of monocotyledonous and dicotyledonous plant cell cultures. Can. J. Bot. 50: 199-204.

Senaratna T, McKersie BD, Bowley SR. (1989) Desiccation tolerance of alfalfa (*Medicago sativa* L.) somatic embryos: Influence of abscisic acid, stress pretreatments and drying rates. Plant Sci. 65: 253-260.

Senaratna T, McKersie BD, Bowley SR. (1990) Artificial seeds of alfalfa (*Medicago sativa* L.) induction of desiccation tolerance in somatic embryos. In Vitro Cell. Dev. Biol. 26: 85-90.

Shahin EA, Spielmann A, Sukhapinda K, Simpson RB, Yashar M. (1986) Transformation of cultivated alfalfa using disarmed *Agrobacterium tumefaciens*. Crop Sci. 26: 1935-1939.

Song J, Sorensen EL, Liang GH. (1990) Direct embryogenesis from single mesophyll protoplats in alfalfa (Medicago sativa L.). Plant Cell Rep. 9: 21-25.

Tabe LM, Wardley-Richardson T, Ceriotti A, Aryan A, McNabb WC, Moore A, Higgins TJV. (1995) A biotechnological approach to improving the nutritive value of alfalfa. J. Plant Sci. 73: 2752-2759.

Tsugawa, H, Otsuky Y, Susuky M. (1998) Efficient transformation of rice protoplasts mediated by a synthetic polycationic amino polymer. Theor. Appl. Genet. 97: 1019-1026.

Tueber LR. and Brick MA. (1988) Morphology and Anatomy p. 125-162 *In* A.A. Hansen (ed.) Alfalfa and Alfalfa Improvement. American Society of Agronomy, Madison, WI.

Wandelt CI, Khan MRI, Craig S, Schroeder HE, Spencer D, Higgins, TJV. (1992) Vicilin with carboxy-terminal KDEL is retained in the endoplasmic reticulum and accumulates to high levels in the leaves of transgenic plants. Plant J. 2: 181-192.

PLANT MOLECULAR FARMING: Using Oleosin Partitioning Technology in Oilseeds

Dr. Maurice M. Moloney

The University of Calgary
Department of Biological Science
2500 University Drive N.W.
Calgary, Alberta T2N 1N4
Canada
Phone: (403)220-6823
Fax: (403)220-0704
Email: mmmolone@acs.ucalgary.ca

INTRODUCTION

The seed, as a means of plant propagation, utilizes sophisticated and versatile storage mechanisms. Seeds store lipids or carbohydrates as a carbon source, protein for carbon and nitrogen, and some inorganics such as phosphate, which is sequestered in phytic acid. The seed is capable of storing these products for extended periods of time often under extreme conditions of cold, heat and water stress. The deposition and storage of proteins occurs in a variety of different tissues (e.g. cotyledons, perisperm, endosperm) within seeds. In addition, the subcellular localization of the stored proteins varies from species to species.

Recognizing that the seed is the site of storage protein deposition in plants leads to questions about which other proteins could be stored in seeds of transgenic plants expressing recombinant proteins. Using gene-transfer techniques, it should be possible to modify the type of proteins in seeds so that they accumulate other proteins, which can be used for therapeutic or industrial purposes.

The concept of using plants as a host for the production of valuable proteins is referred to as "molecular farming" (Goodman *et al.*, 1987). A wide range of proteins and polypeptides have already been expressed in diverse plant tissues and organs, but the features of seeds, which make them stable repositories of stored proteins, provide unique advantages in molecular farming applications.

Plant seeds typically store part of the energy needed for germination in organelles called oil bodies or oleosomes. Oil bodies are spherical structures, comprising an oil droplet of neutral lipid (most frequently triacylglycerides) surrounded by a half-unit phospholipid membrane. A micrograph of typical seed cells containing oil bodies is shown in Figure 1. The surface of these oil bodies is surrounded by a unique class of protein called oleosins. These proteins have an extremely hydrophobic core and

E.E. Hood and J.A. Howard (eds.), Plants as Factories for Protein Production, 55–75.

appear to be lipophilic proteins as reported so far in the protein and DNA databases. Their N- and C- termini are more hydrophilic or amphipathic (Huang *et al.*, 1992). Oleosins become associated with nascent oil bodies during oleosome biogenesis on the ER by a co-translational mechanism (Hills *et al.*, 1993, Loer and Herman, 1993). Oleosins in many oilseeds accumulate at relatively high levels. In the *Brassica* species, for example, oleosins may comprise somewhere between 8-20% of total seed protein (Huang, 1996). This level of accumulation reflects the fact that oleosin genes are strongly transcribed during seed development.

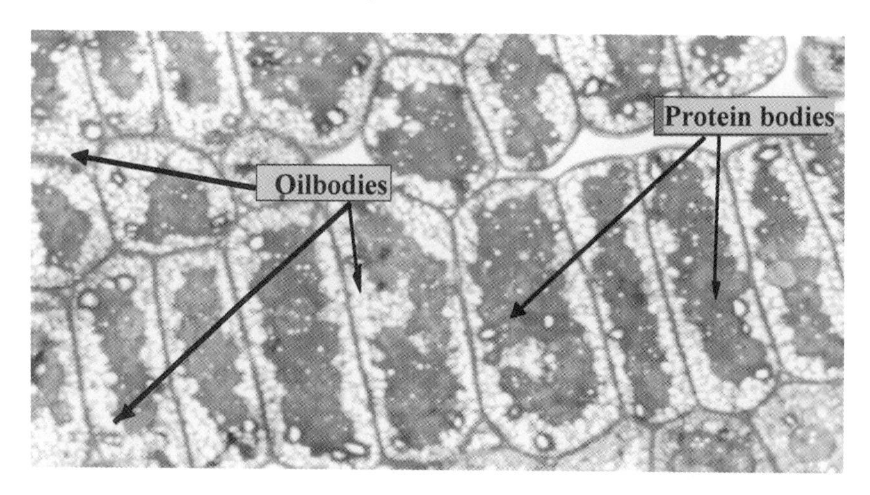

Figure 1. *Transverse section of hypocotyl cells of developing Brassica napus seeds, indicating the protein bodies (dark staining), oil bodies (lighter-stained) spheres and apoplastic space. These subcellular locations have each been shown to act as suitable sites for recombinant protein deposition in seeds*

When oilseeds containing oil bodies are extracted in aqueous solvents, they form a three-phase mixture of insoluble material, aqueous extract and an emulsion of the oil bodies, which on standing or low-speed centrifugation, will result in the flotation of the oil bodies accompanied by their protein complement. Successive aqueous washes of this oil body fraction remove any extraneous proteins bound loosely to oil bodies. The oleosins, conversely, remain tightly associated with the oil bodies due to their highly lipophilic core. Protein analysis of oil body preparations which have undergone only flotation, centrifugation, and washing reveals that virtually all the other seed proteins are absent, thus the oleosin fraction is highly enriched (Figure 2).

Oleosin Profiles in Different Oilseeds

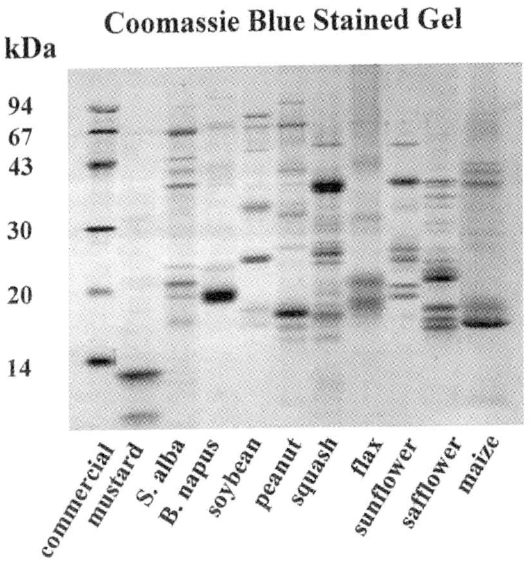

Figure 2. *Coomassie blue stained protein gel of oil body freactions from a variety of common oilseeds. "Commercial mustard" refers to an extract of a comestible mustard preparation ("Grey Poupon") commonly consumed in N. America and indicative of the ubiquity of oleosins in many human food products. The protein patterns of oil body fractions are very simple compared to whole cell extracts.*

These observations have led to the idea that oleosins could serve as vehicles or carriers for heterologous proteins expressed in plant seeds. This would facilitate the expression and simple purification of recombinant proteins in plants. In most protein production systems, a significant proportion of the cost-of-goods is accounted for by the cost of recovery and purification of the protein. Improvements in recovery of a desired recombinant protein could impact the economic viability of the use of plant-based production systems.

Oleosin proteins and their subcellular targeting

Oleosins are near-ubiquitous seed proteins, present in all common oilseeds. They are also constituents of many other seeds not cultivated primarily for their oil, including corn and cotton. In oilseeds such as sunflower (*Helianthus annuus*, safflower (*Carthamus tinctorius*), canola (*Brassica napus*) and flax (*Linum usitatissimum*), triacylglycerol may comprise as much as 40% of dry weight of the seed, and therefore, up to 50% of seed volume may constitute oil bodies. Oil bodies in seeds comprise three components: a neutral lipid lumen, a phospholipid half-unit membrane and a protein "shell" comprising one or more isoforms of the oleosin protein family (Huang, 1992). Oil bodies in seeds occur in relatively large numbers and are typically 0.5-2 microns in diameter. Thus, a cell in a developing *Brassica* embryo may contain hundreds of oil bodies of average size 0.5 microns (as shown in Fig. 1). Other proteins may adhere to oil bodies *in vivo* and, certainly, on extraction from seeds other proteins not usually resident on oil bodies can adhere (Kalinski *et al.* 1992). Despite this it is clear that oleosins associate with oil bodies in a manner very distinct from other cellular proteins. By using relatively harsh treatments such as washing with 8M urea or 0.5M sodium carbonate it is possible to remove all surface-adhering proteins except oleosins from oil bodies. Furthermore, it is now clear that the topology of oleosins is such that their highly lipophilic core is embedded in the lumen of the oil body while the N- and C-termini are cytoplasmically oriented (Huang, 1992; Abell *et al.*, 1997). That oleosins form an apparent "shell" around oil bodies is evident from experiments with phospholipase C, which has no effect on the disruption of oil bodies unless they are pre-treated with a protease such as trypsin. Given the great number of small oil bodies displaying a large surface area in many oilseeds, the oleosin content of many seeds approaches those normally associated with storage proteins. The fundamental role of oleosins appears to be in encapsulating oil-bodies and probably controlling oil-body surface-to-volume ratio, a property, which determines ease of lipolysis of storage lipid during germination. It has also been suggested that oleosins help to prevent coalescence of oil bodies particularly during the early phases of germination (Leprince *et al*, 1998)

Oleosin targeting to oil bodies

The targeting of oleosins to the oil body has been a subject of several studies. Unlike seed storage proteins, oleosins undergo subcellular targeting to oil-bodies without any structural modifications such as cleavage of N- or C-termini (van Rooijen *et al*, 1995a,b). Oleosins appear to use a signal-anchor sequence, which is unaffected by ER lumen proteases. The nascent oleosin polypeptide appears to associate with the ER in a signal recognition particle (SRP) dependent pathway (Abell *et al*, 2002 *in press;* Beaudoin *et al*, 2000). The ER-associated oleosin polypeptide is mobilized to the developing oil bodies by a mechanism that is not yet elucidated. However, it is known that certain structural features such as the presence of a motif known as a "proline knot" is essential for this migration, or at least for the maintenance of oleosin stability on the oil body after transfer (Abell *et al*, 1997). This feature comprises three prolines distributed in a 12 amino acid stretch, which results in a hairpin bend in the polypeptide. It has been suggested that this structural feature also permits the formation of an anti-parallel interaction between the flanking amino-acids of the hydrophobic core, which might impart structural stability to oleosins (Huang, 1992).

Subcellular targeting of recombinant oleosins

While native oleosins are targeted with high avidity to oleosomes *in vivo*, it was not clear whether the addition of polypeptide sequences to oleosins would result in aberrant targeting. Such aberrant targeting has been noted previously with modified storage proteins such as recombinant phaseolin (Hoffman *et al.*, 1988) or 2S albumin (Krebbers and Vendekerckhove, 1990). Consequently, experiments were performed to test the idea that modifications of oleosins at either the N- or C- terminal end might affect overall targeting to oil bodies (Van Rooijen and Moloney, 1995a, Van Rooijen and Moloney, 1995b). In that work, the authors showed that both N- and C- terminal translational fusions of oleosin with ß-glucuronidase (GUS, did not significantly impair the basic targeting mechanism. Furthermore, the C-terminal oleosin-GUS fusion remained enzymatically active. This activity was followed in oil body extracts to test whether the oleosin-GUS protein was attached to the oil body with similar avidity to that of native oleosin. It was found that oleosin fusions targeted and bound to oil bodies with similar avidity to native oleosins.

The domains that determine the targeting of oleosins were investigated by van Rooijen *et al.*, (1995b). In those experiments, it was shown that removal of the highly variable C-terminal of the oleosin protein did not have any significant impact on its subcellular targeting. This was subsequently verified using an in vitro analytical method, which showed that ER association of the oleosin nascent polypeptide is unaffected by removal of the C-terminus of the protein (Abell *et al*, 1997).

These experiments have elucidated a number of important properties of oleosins. First, they can be extended at either end and will still undergo oil body targeting. Second, long polypeptide extensions do not seem to pose a problem (GUS has a molecular weight of ~67 kDa). Furthermore, these experiments were performed using a ß-glucuronidase known to be susceptible to N-glycosylation at position 358. It has been shown that if such a ß-glucuronidase is exposed to the ER lumen, GUS is inactivated due to glycosylation (Iturriaga *et al.*, 1990). In the case of the oleosin C-terminal extensions, GUS was fully functional indicating that it was not glycosylated. This establishes that fusions at the C-terminal of the oleosin are not exposed to the ER lumen and remain on the cytoplasmic side.

As can be seen in Figure 3, an oleosin-fused polypeptide associates predominantly with the oil body and is not substantially associated with other cellular compartments. Upon flotation cetrifugation, the desired protein-fusion is the only protein found in the oil body fraction other than the native oleosin complement.

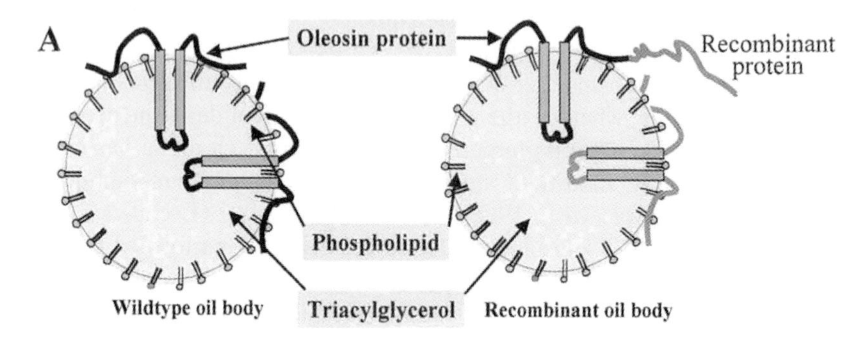

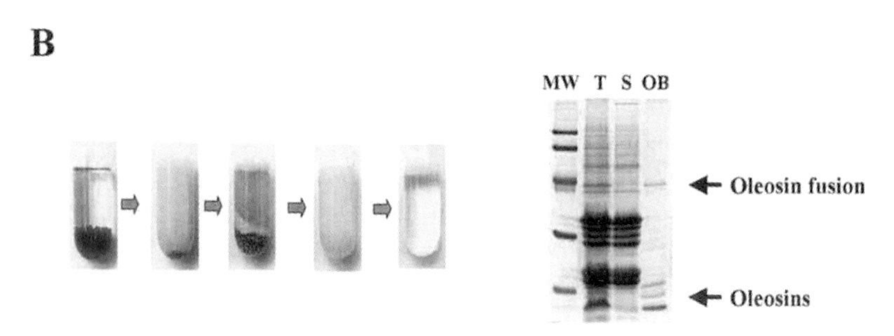

Figure 3A. *Configurations of native and recombinant oleosins on oil bodies. Oleosins comprise a lipophilic central domain, which anchors the oleosin to the oil body. The N- and C-termini of the oleosin are always displayed on the cytoplasmic or aqueous side of the oil body.*
Figure 3B. *Diagram illustrating the purification of oil bodies by flotation centrifugation and the extent of protein enrichment by this process shown in a polyacrylamide gel. MW, molecular weight marker, T=total cell protein, S=soluble protein, OB=oil body-associated protein.*

This subcellular targeting methodology was further refined by interposing a specific labile cleavage site between the oleosin and recombinant protein domain. In the early experiments, these were 4 amino-acid proteolytic sites hydrolysed by enzymes such as thrombin or factor Xa.

Such a configuration should permit the recombinant polypeptide domain to be cleaved from the surface of the oil bodies (See Figure 4). As can be seen from Figure 4 this cleavage can indeed be effected using oil bodies suspended in cleavage buffer and subjected to a specific protease treatment. The resulting product in the aqueous "undernatant" phase is predominantly the desired protein or polypeptide.

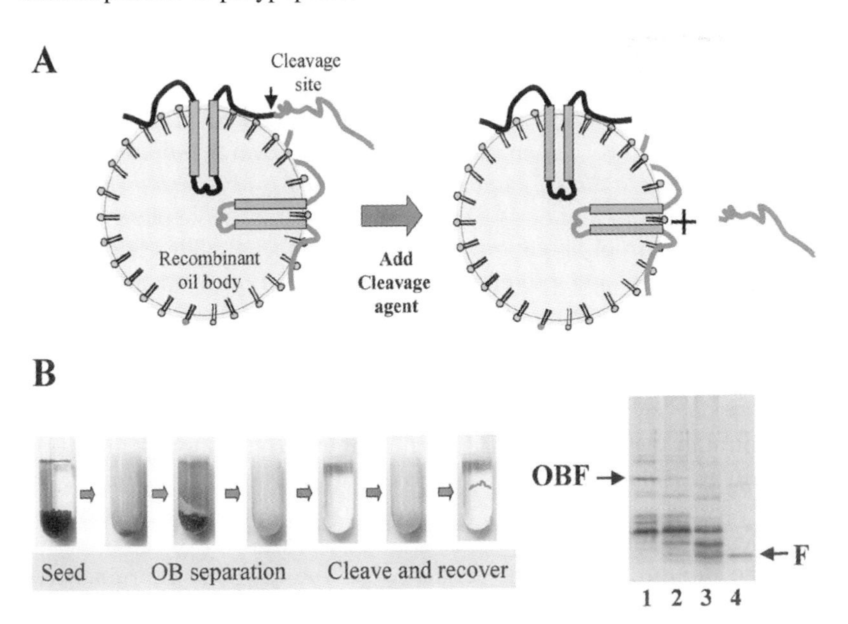

Figure 4 A. Configuration of oleosin fusion proteins and their cleavage from oil bodies.
Figure 4 B. Depiction of the purification and cleavage process and the analysis of the cleavage reaction. 1. Oil body preparation of recombinant oil bodies, 2. Treatment of oil bodies with 1:1000 cleavage enzyme to fusion protein, 3. Treatment of oil bodies with 1:100 cleavage enzyme to fusion protein, 4. Aqueous "undernatant" fraction after cleavage. OBF = oil body fusion; F = cleaved target protein.

Using oleosins for the production of the anti-coagulant, hirudin

While the above experiments demonstrate the basic principles of oil body-based recombinant protein production, it is of great interest to apply it in cases where the recombinant protein is of high-value. This would test the economics as well as the technical parameters of the system. To this end, we created a translational fusion between an *Arabidopsis* oleosin coding sequence and the coding sequence of the mature form of the blood anti-coagulant, Hirudin. Hirudin is a naturally occurring thrombin inhibitor produced and secreted in the salivary glands of the medicinal leech, *Hirudo medicinalis.* Hirudin is a powerful anticoagulant with a number of very desirable properties including stoichiometric inhibition of thrombin, short clearing time from the blood and low immunogenicity. The unit cost of production in leeches would be prohibitive for extensive therapeutic applications. Hirudin has been made in a variety of microorganisms including *E. coli* and yeast, but these entail significant fixed costs associated with fermentation. Our objective, therefore, was to test the oleosin expression and purification system as an alternative and potentially inexpensive source for hirudin.

Constructs containing the translational fusion of oleosin-hirudin under the transcriptional control of an oleosin promoter from *Arabidopsis* were introduced into *Brassica napus* using *Agrobacterium*-mediated transformation (Moloney *et al.*, 1989). The resulting transgenic plants yielded seed in which a protein of about 25 kDa could be detected. This is shown in Figure 5, where the fusion protein is visualized both with an anti-hirudin antibody and an anti-oleosin antibody. The fusion protein cross-reacted efficiently with a monoclonal antibody raised against hirudin. This protein proved to be associated tightly with the oil bodies and could not be removed from the oil bodies by successive washings. Attempts to determine whether the oleosin-hirudin fusion protein had antithrombin activity suggested that the fusion protein was completely inactive. As part of the construction of the translational fusion, we had incorporated four additional codons specifying a Factor Xa cleavage site (I-E-G-R). This configuration was designed to permit release of the hirudin polypeptide sequence off the oil body by proteolytic cleavage. When the washed oil bodies from these seeds were treated with Factor Xa, hirudin polypeptide was released into the aqueous phase. As can be seen in Figure 6, only the cleaved hirudin product showed this anti-thrombin activity. It is noteworthy that no refolding steps were required to reveal this activity. Authentic hirudin has three disulfide bridges. These are essential to its activity. Thus, if the plant were to make hirudin polypeptide, but did not allow its correct folding no thrombin inhibition would be detected unless a refolding treatment was applied. In fact, after Factor Xa treatment, the aqueous phase showed strong antithrombin activity, demonstrating that biologically-active hirudin was released. The specific activity of the inhibitor was similar to that of recombinant hirudin secreted from yeast cells, indicating that the majority of the hirudin released was correctly folded and disulfide bridges were appropriately configured. The released hirudin was then subjected to ion-exchange chromatography (Mono Q). The fractions showing anti-thrombin activity were concentrated and loaded onto a C_{18} reversed phase analytical HPLC column. This showed that the hirudin was substantially pure after the ion-exchange step (Parmenter *et al*, 1995;

Parmenter *et al*, 1996). The HPLC trace showed two peaks close to the expected retention volume of hirudin. Mass spectrometric analyses of these two peaks using TOF-MALDI revealed that the two peaks were full-length hirudin and a truncated product from which two C-terminal amino-acids were missing. Interestingly, such a truncated form was also found when Hirudin was expressed in yeast (Heim *et al.*, 1994). Both forms of the molecule are potent thrombin inhibitors.

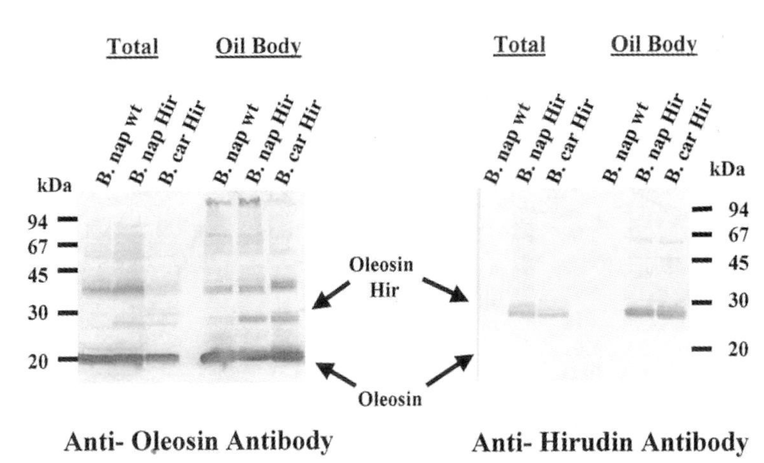

Figure 5. *Western blot analysis of oleosin-hirudin fusions in seeds of Brassica napus (B. nap) and Brassica carinata (B. car). Left hand panel was visualized using anti-oleosin antibodies, right hand panel was visualized using anti-hirudin antibodies.*

Hirudin anti-coagulant activity in transgenic seeds

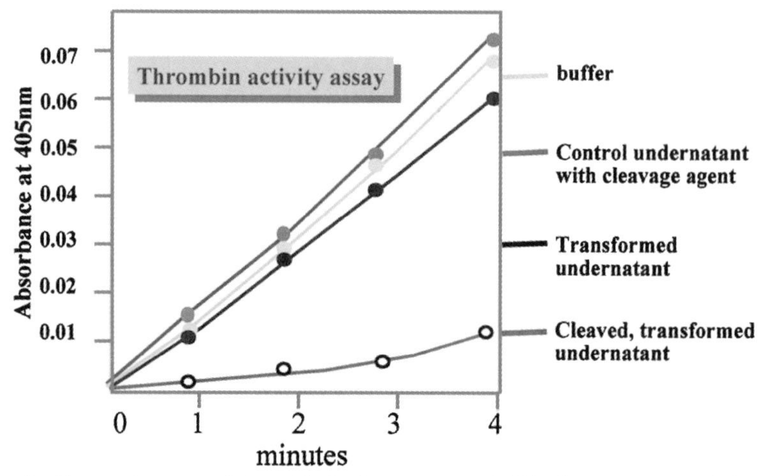

Figure 6. *Enzymatic assay of thrombin in the presence of oil body extracts containing or not containing hirudin. Increases in the absorbance at 405 nm indicate cleavage of substrate by thrombin. In this assay, only cleaved oleosin-hirudin givers rise to an active anti-thrombin product.*

A number of improvements would be required to render this system economical. These are a) expression levels, b) cleavage efficiency and c) cleavage system (use of Factor Xa exemplifies the system but would be prohibitively expensive). Expression levels of the recombinant protein have in a few cases been about one-tenth of the native oleosin accumulation. This corresponds to about 1% of total cellular protein. It is, however, expected that using stronger promoters than the native oleosin promoters, higher levels of accumulation will be sustained. In fact, substitution in these constructs with stronger promoters such as that from phaseolin, routinely yields protein expression levels in excess of 2% of total cell protein. This is well beyond the economic threshold for most recombinant proteins in this system. Cleavage efficiency varies from one fusion protein to another and may be affected by the conformation of the target protein but also by the spacing between the cleavage site and the oil body surface. Finally, the use of Factor Xa here was essentially as a convenient experimental system. In practice it would be essential to have an inexpensive cleavage enzyme or chemical to render this process competitive. Recent experiments with inexpensive enzymes such as chymosin suggest that alternative cleavage systems are compatible with this process and would allow the cost of the cleavage step to be minimized (Moloney *et al*, 1998).

Oil body fusions for the production of immobilized enzymes

The potential for using oleosin fusions as an adaptable means of recombinant protein expression in seeds is under investigation. It is interesting to note that the production of oleosin-polypeptide fusions appears to function for rather short peptides such as IL-1ß and Hirudin, but equally well for much longer polypeptides such as ß-glucuronidase. In cases where larger polypeptides are produced there is good evidence that proper folding occurs. This is supported by the fact that in a number of cases including ß-glucuronidase and xylanase, the fusion protein retains its enzymatic activity. This finding has also led to experiments which illustrate the utility of oil bodies as immobilization matrices (Kuhnel *et al.*, 1996, van Rooijen and Moloney, 1995b). Using ß-glucuronidase as a model it was shown that a dispersion of oil bodies carrying ß-glucuronidase on their surface would hydrolyse glucuronide substrates for GUS with catalytic properties indistinguishable from soluble GUS (Kuhnel *et al*, 1996). Furthermore, it was shown that virtually all the enzymatic activity could be recovered and recycled by the use of flotation centrifugation to obtain the oil bodies. The enzyme in dry seed remains fully active for several years. Once extracted onto oil bodies the half-life of the enzyme was about 4 weeks (van Rooijen and Moloney, 1995b).

The expression of the enzyme, xylanase, on oil bodies illustrates several ways in which the basic technology could be used (Liu *et al*, 1997). The enzyme accumulated on seed oil bodies in active form. From here it can be cleaved to release soluble xylanase into solution. In these experiments it was shown that the immobilized xylanase had essentially the same kinetic properties as soluble xylanase from the same source (*Neocallimastix patriciarum*). Such a preparation could be useful in such processes as de-inking of recycled paper. Alternatively, without significant purification, the oil bodies could be mixed with insoluble substrate such as crude wood pulp to help break down the xylan crosslinks. This could be helpful in paper-making. Cost would be minimised as the enzyme itself could be recovered and re-used. In addition to these formulations the enzyme could be delivered in seed meal to animals without purification. In this embodiment, the enzyme would function in the stomach of a monogastric animal to improve digestibility of meal. Clearly, this latter formulation would be created in conjunction with other cellulytic enzymes. The result, however, would be to provide monogastrics with enhanced feed efficiency and reducing biological waste.

Expression of somatotropins as oleosin fusions - oil bodies as delivery vehicles

As is demonstrated by work on xylanase and GUS, oleosin fusions may remain active without cleavage from the oil body. This may result in a novel formulation for the delivery of a biologically active molecule. This is exemplified by the expression of a cyprinid (carp) somatotropin as an N-terminal oleosin fusion (Mahmoud, 1999). It has proven difficult to express somatotropins in plants at economically viable levels (Bosch *et al*, 1994). Oleosins appear to offer several advantages for the accumulation and use of this class of protein. In this experiment, the coding sequence of growth hormone from the common carp was fused in frame with an oleosin coding sequence. This was used to create transgenic *Brassica napus* plants expressing the fusion in a seed-specific manner. The resulting fusion also contained a cleavage site so that it could be released from the oil body by endoproteolysis. The resulting polypeptide was of the correct molecular weight and showed biological activity when injected into goldfish as determined by the stimulation of insulin-like growth factor-1 (IGF-1) transcription. Fish somatotropins are also known to be active orally (Jeh *et al*, 1998), however, it was not clear whether the fusion polypeptide would show oral activity. To test this, formulations of feed pellets were prepared into which were incorporated oil body preparations containing the oleosin-somatotropin fusion protein. It was shown with rainbow trout that such feeding experiments resulted in substantial, statistically significant increases in growth rates over an 8 week period. Results from one of these feeding trials are shown in Figure 7. These results further demonstrate the use of oil bodies not only for protein expression and recovery, but also for their use as delivery vehicles and as media for formulation of orally delivered biologically-active molecules.

Growth Response in Rainbow Trout to Oleosin cGH (oral delivery)

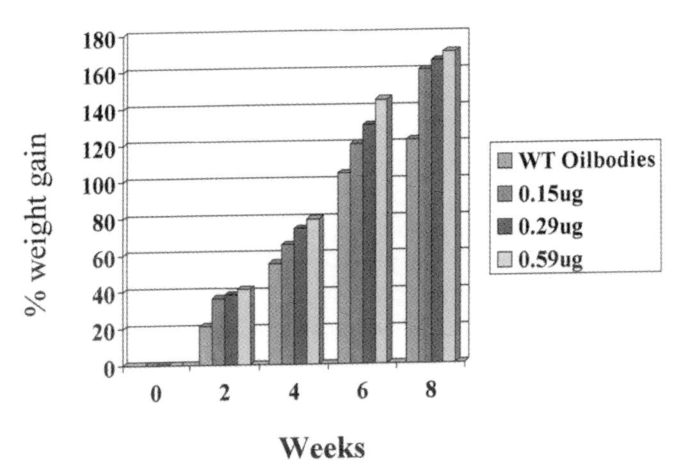

Figure 7. *Effect of orally delivered oleosin-carp somatotropin as a fusion protein on oil bodies which were incorporated directly into feed. The figures 0.15, 0.29, 0.59ug are incorporation rates in micrograms per gram of body weight of the fish at the start of the experiment.*

Recovery of recombinant proteins from seeds

Besides these benefits, which favour seeds as vehicles for recombinant-protein production, seeds also lend themselves readily to the recovery and purification of the transgene product. To date there have been only a few studies performed on recovery of recombinant proteins from seeds (Parmenter *et al.*, 1996; Hood *et al.*, 1997; Kusnadi *et al.*, 1998a,b). A significant advantage of seeds as hosts for recombinant proteins is that seed processing is a very sophisticated industry. In consequence, there is a plethora of existing processing equipment and technology available.

Aqueous extraction systems

Where no purification is required a technique as simple as dry milling may be adequate for the production of a recombinant protein formulation. This was the approach used by Pen *et al.* (1993) for phytase. Normally, however, it will be essential to extract and at least partially purify the desired protein. Where seeds contain substantial quantities of oil, it is customary to employ hexane extraction to recover maximal amounts of the oil (see Figure 8a). In most instances involving recombinant proteins, this step is not possible due to the denaturant property of hexane and many organic solvents. Therefore, extraction and purification of recombinant seeds will most likely require aqueous extraction (see Figure 8b). This has been performed successfully at scale by Kusnadi *et al.* (1998a,b) for the extraction of avidin and ß-glucuronidase (GUS) from corn seed. They noted that in spite of using a constitutive (i.e. non tissue-specific) promoter up to 98% of the GUS activity was deposited in the germ (embryo). This observation led to an advantageous extraction scheme in which the kernel separated into the endosperm and the embryo. The full-fat germ was then extracted using aqueous buffers to obtain a protein fraction, or alternatively the germ was first defatted with hexane at 60°C. Surprisingly, the GUS enzyme was not inactivated by this solvent treatment, although this may be a property peculiar to GUS which is generally considered to be a very stable enzyme. The aqueous protein-rich fraction was subjected to four rounds of chromatography including two ion exchanges, one hydrophobic interaction and one size-exclusion step. The overall yield of purified protein was 10%. This purification scheme would be quite costly on a large scale and with such a low recovery would only be economical for high-value proteins. Nevertheless, the scheme is simple and lends itself readily to optimization of each individual step of the process.

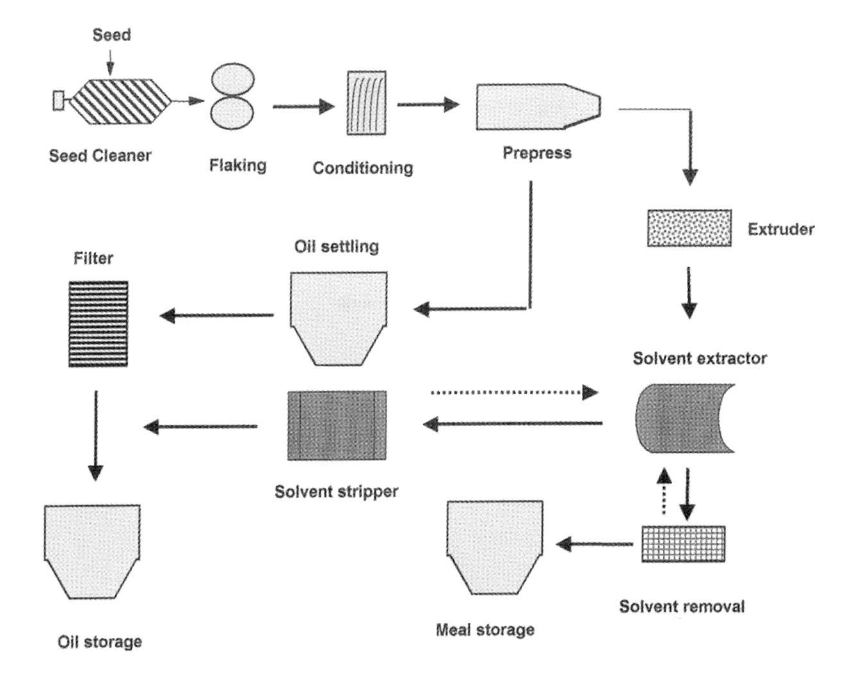

Figure 8A. *Process diagram depicting traditional extraction of oleaginous seeds which typically involves a heating step (conditioning) and hexane extraction. Without significant modification, it would be difficult to extract biologically active proteins using this traditional procedure.*

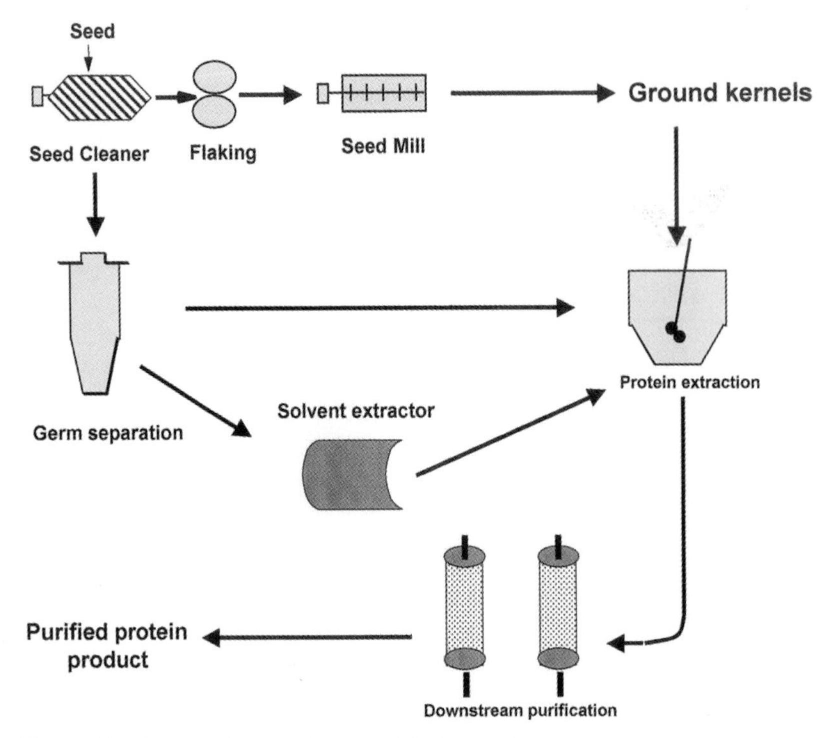

Figure 8B. *Process diagram for a modified procedure to permit extraction of recombinant proteins from corn while recovering some traditional by-products (after Kusnadi et al., 1998 a, b).*

Two-phase extraction systems using oil bodies

An alternative extraction scheme for separation and purification to homogeneity of a protein is based on oil body partitioning (Parmenter *et al.*, 1996). In this system the host-seed that has been used is canola, and thus apart from dehulling, little pre-fractionation is possible. However, aqueous extraction of the whole seed followed by centrifugation to separate oil bodies has proven to be a major enrichment step. This alternative process is depicted in Figure 9. Once the oil bodies are washed only minor amounts of other seed proteins remain. In the case of hirudin for example, the cleavage of recombinant oleosin-fusion protein from oil bodies provided a significant enrichment step. This extract was subjected to anion and reverse-phase chromatography and yielded an extraction efficiency of purified hirudin of about 40% with a purity in excess of 99% (Parmenter *et al*, 1996).

Minimization of chromatography steps and early-stage enrichment of the desired protein assist greatly in overall rates of recovery.

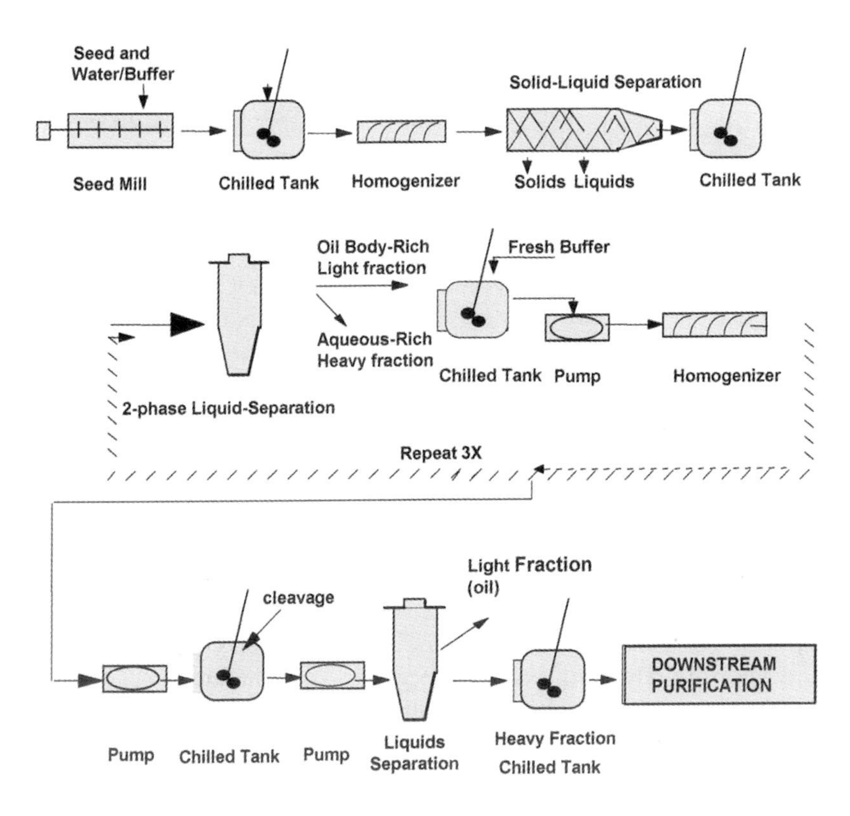

Figure 9. *Process diagram showing a two-phase partitioning system used with aqueous extraction to recover recombinant proteins from oilseeds particularly where the desired protein is associated with oil bodies or oleosins.*

Recovery and purification of expressed recombinant proteins from transgenic plants is probably the most critical factor in establishing plants as a practical alternative system for protein production. While a variety of schemes might be envisaged, it is essential that the number of processing steps be minimized and that each step be carried out at much higher efficiencies than have thus far been reported. This is, however, an area that has only recently received much attention and it is clear that greater efficiencies will be forthcoming.

Recovery of recombinant proteins from oil bodies

In the initial work leading to the use of oleosins as carriers of recombinant proteins, the desired protein must be expressed as a translational fusion. This poses the problem of cleavage of the recombinant protein from the oil body. Although there are many examples of cleavage enzymes that can be used at a laboratory scale such as Factor Xa, thrombin and enterokinase, it is generally accepted that these enzymes are too expensive for larger scale use and often do not provide a high efficiency of cleavage. These observations led us to consider alternative approaches to cleavage reactions, using examples of high efficiency cleavage in nature. A large number of proteins undergo specific cleavage reactions as part of their post-translational processing. This includes the cleavage of signal sequences during subcellular targeting and the maturation of a number of proteases. The cleavage reactions of some of the self-maturing proteases have been widely studied. However, it appears that little work has been performed on understanding their ability to function *in trans* with a heterologous polypeptide. We therefore investigated this, using as a model maturation reaction, the self-processing of the milk-clotting enzyme chymosin. In this work, we examined the potential of chymosin to function as a cleavage enzyme of a recombinant fusion protein which comprised the chymosin A propeptide leader fused translationally to a number of heterologous proteins. Using this configuration, it was shown that chymosin can act as a cleavage enzyme for a heterologous polypeptide associated with the chymosin pro-sequence. In fact, the cleavage reaction appears to be highly efficient and can result in almost stoichometric reactions (Moloney *et al*, 1998).

This approach provides for broad applications not only with fusion proteins associated with oil bodies, but also with fusion proteins produced by more conventional routes such as fermentation. Clearly, given the efficiency of the reaction and the relative cost of chymosin rather than Factor Xa, this process holds much promise for commercial scale preparative procedures. A full report of this cleavage system and its uses will be published elsewhere (Kuhnel *et al*, 2002, submitted)

Use of Oil Bodies as Affinity Matrices: non-covalent polypeptide attachment

In the original conception of oleosins as carriers of recombinant proteins, the desired polypeptide was expressed as a covalently-associated fusion protein as described above. Nevertheless, it is clear that there are many polypeptides, which may require exposure to a subcellular compartment other than the cytoplasm, where oleosins are effectively displayed. For example, it may be necessary for certain proteins to be exposed to the lumen of the ER where appropriate disulfide bridging or other post-translational modifications can occur. It was, therefore, reasoned that such proteins deposited into other cellular compartments might be recovered by the association of the polypeptide with oil bodies using non-covalent binding. An example of this would be the inclusion of an "affinity tag" on the recombinant polypeptide so that when it is exposed to the oil body during processing it preferentially associates with the oil body fraction. This would permit the recovery of such proteins using the same process of flotation centrifugation currently used for covalently attached polypeptides. An example of this is the use of a single-chain antibody (scFv) against oleosin which has a high affinity for the oil bodies. This scFv can be expressed as a fusion protein with other polypeptides. In the presence of oil bodies carrying the recognized oleosin, the scFv adheres with high avidity to the oil bodies and thus allows for oil body washing and protein recovery. This method (Moloney *et al*, 1999) has broad applicability and allows for the recovery of proteins deposited in cellular compartments other than the oil body itself.

Conclusions and future prospects

The use of plant oil bodies and their associated proteins, oleosins as vehicles for recombinant protein product has been illustrated with a number of examples. The major advantage of using oil bodies as carriers is the ease with which proteins can be recovered and purified. Oleosin targeting does not seem to be impaired even by very long polypeptide extensions to the N- or C-termini of the oleosin. This greatly enhances the versatility of this system in contrast to alternative approaches for recombinant protein production in plants. Separation of oil bodies from seed extracts is amenable to scale-up using equipment typical to dairy operations such as cream-separators (Jacks *et al.*, 1990). In consequence, it seems likely that this system could be used for the production of a wide range of proteins of therapeutic, industrial and feed or food use. We are investigating production of a wide range of commercially attractive polypeptides in this system and have developed scaled-up extraction and purification systems, which could be applied to a variety of different oilseeds engineered for production of such oleosin-polypeptide fusions.

REFERENCES

Abell BM, Holbrook LA, Abenes M, Murphy DJ, Hills MJ, and Moloney MM. 1997. Role of the proline knot motif in oleosin endoplasmic reticulum topology and oil body targeting. Plant Cell 9:1481-93

Beaudoin F., Wilkinson BM, Stirling CJ, and Napier JA. 2000. In vivo targeting of a sunflower oil body protein in yeast secretory (sec) mutants. Plant J. 23 :159-70

Bosch D, Smal J, and Krebbers E. 1994. A trout growth hormone is expressed, correctly folded and partially glycosylated in the leaves but not the seeds of transgenic plants. Transgenic Research 3: 304-310

Goodman RM, Knauf VC, Houck CM, and Comai L. 1987. Molecular Farming. European Patent Application. WO 87/00865, PCT/US86/01599

Heim J, Takabayashi K, Meyhack B, Maerki W., Pohlig G. 1994. C-terminal proteolytic degradation of recombinant desulfato-hirudin and its mutants in the yeast *Saccharomyces cerevisiae.* Eur. J. Biochem. 226, 341-353

Hills MJ, Watson MD, and Murphy DJ. 1993. Targeting of oleosins to the oil bodies of oilseed rape (*Brassica napus L.*). Planta 189, 24-29

Hoffman LM, Donaldson DD, and Herman EM. 1988. A modified storage protein is synthesized, processed and degraded in the seeds of transgenic plants. Plant Mol. Biol. 11, 717-729

Hood EE, Witcher DR, Maddock S, Meyer T, Baszczynski C, Bailey M, Flynn P, Register J, Marshall L, Bond D, Kulisek E, Kusnadi A, Evangelista R, Nikolov Z, Wooge C, Mehigh RJ, Hernan R, Kappel WK, Ritland D, Li CP and Howard JA. 1997. Commercial production of avidin from transgenic maize: characterization of transformant, production, processing, extraction and purification. Molecular Breeding 3: 291-306

Huang AHC. 1992. Oil bodies and oleosins in seeds. *Annual Reviews of Plant Physiology and* Plant Molecular Biology 43: 177-200

Huang AHC. 1996. Oleosins and oil bodies in seeds and other organs. Plant Physiol. 110, 1055-1061

Iturriaga G, Jefferson RA, and Bevan MW. 1990. Endoplasmic reticulum targeting and glycosylation of hybrid proteins in transgenic tobacco. Plant Cell 1, 381-390

Jacks TJ, Hensarling TP, Neucere JN, Yatsu LY, and Barker RH. 1990. Isolation and physiocochemical characterization of the half-unit membranes of oilseed lipid bodies. J.A.O.C.S. 67, 353-361

Jeh HS, Kim CH, Lee HK, and Han K. 1998. Recombinant flounder growth hormone from *Escherichia coli*: overexpression, efficient recovery, and growth-promoting effect on juvenile flounder by oral administration. J Biotechnol. 60 183-93.

Kalinski A, Melroy DL, Dwivedi RS, and Herman EM. 1992. A soybean vacuolar protein related to thiol proteases is synthesized as a glycoprotein precursor during seed maturation. Journal of Biological Chemistry 267: 12068-12076.

Krebbers E, and Vandekerckhove J. 1990. Production of peptides in plant seeds. TIBTECH 8, 1-3

Kühnel B, Holbrook LA, Moloney MM, and Van Rooijen GJH. 1996. Oil bodies of transgenic *Brassica napus* as a source of immobilized ß-glucuronidase. J.A.O.C.S. 73, 1533-1538

Kusnadi AR, Hood EE, Witcher DR, Howard JA, and Nikolov ZL. 1998a. Production and purification of two recombinant proteins from transgenic corn. Biotechnology Progress 14: 149-155

Kusnadi AR, Evangelista RL, Hood EE, Howard JA, and Nikolov ZL. 1998b. Processing of transgenic corn seed and its effect on the recovery of recombinant beta-glucuronidase. Biotechnology and Bioengineering 60: 44-52

Liu J-H, Selinger LB, Cheng K-J, Beauchemin KA, and Moloney MM. 1997. Plant seed oil-bodies as an immobilization matrix for a recombinant xylanase from the rumen fungus *Neocallimastix patriciarum.* Molecular Breeding 3: 463-470

Loer D, and Herman EM. 1993. Cotranslational integration of soybean (Glycine max) oil body membrane protein oleosin into microsomal membranes. Plant Physiol. 101, 993-998

Moloney MM, Walker JM, and Sharma KK. 1989. High efficiency transformation of *Brassica napus* using *Agrobacterium* vectors. Plant Cell Reports 8, 238-242

Moloney MM, Alcantara J, van Rooijen GJH. 1998. Method for Cleavage of Fusion Proteins PCT Patent Application # WO 98/49326

Parmenter DL, Boothe JG, van Rooijen GJH, Yeung EC, and Moloney MM. 1995. Production of biologically active hirudin in plant seeds using oleosin partitioning. Plant Molecular Biology 29: 1167-1180

Parmenter DL, Boothe JG, and Moloney MM. 1996. Production and purification of recombinant hirudin from plant seeds, in MRL Owen and J Pen (eds.), Transgenic plants: a production system for industrial and pharmaceutical proteins, John Wiley and Sons Ltd., pp 261-280

Pen J, Verwoerd TC, van Paridon PA, Beudeker RF, van den Elzen PJM, Geerse K, van der Klis JD, Versteegh HAJ, van Ooyen AJJ, and Hoekema A. 1993. Phytase-containing transgenic seeds as a novel feed additive for improved phosphorus utilization. Bio/Technology 11: 811-814

van Rooijen GJH, and Moloney MM. 1995a. Structural requirements of oleosin domains for subcellular targeting to the oil body. Plant Physiol. 109, 1353-1361

van Rooijen GJH, and Moloney MM. 1995b. Plant seed oil-bodies as carriers for foreign proteins. Bio/Technology 13, 72-77

Part II
Recombinant Protein Products From Plants

HUMAN PHARMACEUTICALS PRODUCED IN PLANTS

James W. Larrick MD PhD, Lloyd Yu PhD, Clarissa Naftzger PhD, Sudhir Jaiswal PhD and Keith Wycoff PhD

Planet Biotechnology, Inc.
2438 Wyandotte Street
Mountain View, CA 94043, USA

Palo Alto Institute of Molecular Medicine
2462 Wyandotte Street
Mountain View, CA 94043, USA

ABSTRACT

Genetically engineered agricultural plants with improved traits (e.g. herbicide and pest resistance) have yielded significant agricultural revenues over the past 5 years. In addition, numerous immunotherapeutic proteins, antibodies and vaccines have been successfully produced using plant "bioreactors"(Arakawa *et al.,* 1998); however, only a limited number have made their way into clinical trials. The most advanced product in human clinical trials is a secretory IgA antibody comprised of four polypeptide chains that inhibits the binding to teeth of *Streptococcus mutans*, the major causal agent of tooth decay. This chapter will summarize recent work demonstrating the potential of plants to synthesize and assemble complex proteins suitable for human therapeutic use.

HISTORICAL BACKGROUND

During the past 100 years antibodies have evolved into a remarkable platform technology for generating therapeutic molecules to benefit human and animal health. The recognized potential of antibody therapy was evident early this century when von Behring and Kitasato (1890) received the first Nobel Prize for their demonstration that passive administration of immune sera could prevent or treat certain infectious diseases. Based on this

E.E. Hood and J.A. Howard (eds.), Plants as Factories for Protein Production, 79–101.

pioneering work passive "serum immunotherapy" using rabbit or horse immune sera was developed and widely used for the treatment of pneumococcal pneumonia, meningococcal meningitis, diphtheria, scarlet fever, measles, tuberculosis, etc. (Cassadeval *et al.*, 1994, 1995, 1997). Although serum therapy was remarkably effective for certain conditions, a number of problems, primarily serum sickness, limited the extent of its use. With the advent of antibiotics in the 1930s (sulfonamides - 1935) and 1940s (penicillin - 1942) the use of immune serum was largely abandoned. However, toxin-mediated diseases such as tetanus, botulism, diphtheria and snakebites as well as digitalis drug toxicity still continue to be treated with polyclonal antibody preparations.

Despite the recognized potential and usefulness of antibodies, the modern renaissance in antibody therapy awaited the development of monoclonal antibodies and recombinant DNA technology (Gavilondo and Larrick, 2000). The first therapeutic antibody product approved by the FDA was the murine monoclonal antibody Orthoclone-OKT3 in 1986. Since then a number of major antibody-based products have been approved: ReoPRo in 1994, Rituxan and Zenapax in 1997, Simulect, Synagis, Remicade and Herceptin in 1998 and Mylotarg (the first antibody drug conjugate) in 2000. Several other major products are scheduled for approval within the next year.

Based on the success of these products and advances in technology, antibodies now account for the single largest group of biotechnology-derived molecules in clinical trials, with a prospective market reaching several billion dollars. According to the Pharmaceutical Research and Manufacturers of America (see http://www.phrma.org), 20% of all biopharmaceuticals in clinical development in 1998 were antibody products. One commissioned global survey indicated that as many as 264 companies are working on over 700 therapeutic antibodies, 220 of which are in clinical trials. The interest in antibody-based therapeutics is a direct consequence of the introduction of genetically engineered immunoglobulins (Borrebaeck, 1995, de Haard *et al.*, 1998), and the refinement of targets for antibody therapy (Gavilondo and Larrick, 2000).

Although these advances have generated an entire industry of novel therapeutic proteins, cost efficient production (Table 1) remains a major issue in the commercialization of therapeutic antibodies. Polyclonal antibody preparations from human blood donors are still valued at over 500 million dollars despite problems with contamination by viruses. Mammary tissue-specific promoters can direct antibody production with correct assembly and function of the foreign antibodies in milk of transgenic animals. Genzyme Transgenics (Framingham, MA; http://www.genzyme.com/transgenics) has used this approach to make chimeric and humanized antibodies for clinical trials

(Pollock *et al.*, 1999). Production of antibodies in milk suffers from lengthy development timelines, and viral or prion contamination remain issues. Efforts to develop bovine colostrum have met with limited success. As described elsewhere in this volume and below, among the newer methods, plants appear most promising for large-scale, efficient, viral-free production of pharmaceutical proteins.

Table 1: *Sources of Therapeutic Antibodies*

Source	Stage	Limitations/advantages
Human plasma	Marketed & in development	Non-specific; expensive, viral risk, IgG only; polyclonal
Human plasma, immune	Marketed & in development	Polyclonal; expensive, viral risk, IgG only
Animal cells, monoclonals	Marketed & in development	Highly specific; expensive, viral risk, IgG and IgM only
Bovine colostrum	In development	Non-specific, IgG/IgA, oral use only
Transgenic animals	In development	Monoclonal; prion/viral risk, long development time
Transgenic plants	In development	Monoclonal; cost effective, safe, secretory IgA possible

TRANSGENIC PLANT BIOREACTORS

It has been almost 20 years since the first generation of transgenic plants (Fraley *et al.*, 1983; Zambryski *et al.*, 1983). Since then, many recombinant proteins have been expressed in several important agronomic species of plants (Fischer *et al.,* 1999a, b) including tobacco, corn, tomato, potato, banana, alfalfa (Austin, 1994), and canola (summarized in Kusnadi *et al.*, 1997; Hood and Jilka, 1999). Recent work suggests that plants will be a facile and economic bioreactor for large-scale production of industrial and pharmaceutical recombinant proteins (Kusnadi *et al.*, 1997; Austin *et al.*, 1994; Krebbers *et al.*, 1992; Whitelam *et al.*, 1994; Larrick *et al.,* 1998). Plants have numerous advantages as production factories for proteins compared with human or animal fluids/tissues, recombinant microbes, transfected animal cell lines, or transgenic animals. Among these are:

- Rapid scale-up of production
- Facile, genuine assembly of multimeric antibodies (unlike bacteria)
- Increased safety, because plants do not serve as hosts for human pathogens, such as HIV, prions, hepatitis viruses, etc.
- Low cost production of bulk crude material on an agricultural scale

suitable for further GMP purification

- Capitalization costs of manufacturing that are perhaps 10% relative to steel tank bioreactor methods

Depending upon the promoters used, expressed transgenic proteins are deposited throughout the plant or in specific parts of the plants (e.g. seeds) or specific organelles within a given plant cell. Much of the early work utilized generic promoters, such as the cauliflower mosaic virus 35S promoter giving protein expression in all parts of the green biomass. De Wilde *et al.* (1998) showed that, after secretion from cells in *Arabidopsis* plants, antibody or Fab fragments accumulate at the sites where water passes on its radial pathway towards and within the vascular bundle. A large proportion of these proteins are transported in the apoplast of *A. thaliana*, possibly by the water flow in the transpiration stream (Conrad and Fiedler, 1998). In contrast, numerous labs have shown transgenic protein accumulation in seeds of tobacco (Fiedler and Conrad, 1995) corn (Russell, 1999; Hood and Jilka, 1999) soybean (Zeitlin *et al.*, 1998), barley (Horvath *et al.*, 2000), rice and wheat (Stoger *et al.*, 2000). Details of some of these studies are presented elsewhere in this volume.

Transgenic plants have been successfully used as bioreactors to produce numerous proteins of pharmaceutical value. Table 3 lists some examples of proteins produced in transgenic tobacco. The diversity of proteins produced is remarkable. The plants cells can synthesize and assembly complex proteins such as triple helicial collagen Ruggiero *et al.*, 2000) and secretory IgA which is comprise of four different peptide chains (see below). In some cases the production levels are unremarkable, probably because codon-optimized constructs were not used. In other cases, such as antibodies (see below) the expression levels are up to several percent of total soluble protein. Based on the first decade of work with transgenic plant expression of pharmaceutical proteins, it is likely that one or more of these will be successfully commericialized in the second decade of work. The most likely of these are antibodies as discussed next.

Table 2. *Antibody-derived molecules produced in transgenic plants*

	Antigen	Plant species-Comment	Reference
Single domain (dAb) (1)	Substance P (neuropeptide)	*Nicotiana*	Benvenuto *et al.* (1991).
Single chain Fv (scFv) (1)	Phytochrome	*Nicotiana*	Firek *et al.* (1993), Owen *et al.* (1992).
ScFv (1)	Artichoke mottled crinkle virus coat	*Nicotiana*--viral protection	Tavladoraki *et al.* (1993 and 1999).
ScFv (1)	Abscisic acid	*Nicotiana*--wilty phenotype	Artsaenko *et al.* (1995); Phillips *et al.*. (1997).
ScFv (1)	Root-knot nematode	*Nicotiana*	Rosso *et al.* (1996).
ScFv (1)	Beet necrotic yellow vein virus	*Nicotiana benthamiana*	Fecker *et al.* (1996, 1997)
ScFv (1)	CEA	Wheat/rice/ tobacco	Stoger *et al.* (2000), Vaquero *et al.* (1999).
ScFv (1)	Several	*Nicotiana* KDEL augments expression	Fiedler *et al.* (1997); Schouten *et al.* (1996, 1997); Bruyns *et al.* (1996).
ScFv (1)	n.a. This scFv is from a lymphoma and used as vaccine to stimulate anti-Idiotype antibodies	*Nicotiana* Transient expression using a TMV vector	McCormick *et al.* (1999)
Fab; IgG (k) (2)	Human creatine kinase	*Nicotiana Arabidopsis*	de Neve *et al.* (1993).
IgG (k) (2)	Transition-state analogue	*Nicotiana*	Hiatt *et al.* (1989); Hein *et al.* (1991).
IgG (k) (2)	Fungal cutinase	*Nicotiana*	van Engelen *et al.* (1994).
IgG (2)	TMV	*Nicotiana*	Schillberg *et al.* 1999.
IgG (k) and IgG/A hybrids (2)	*Streptococcus mutans* adhesin	*Nicotiana*	Ma *et al.* (1994).
Secretory IgA/G (4)	*Streptococcus mutans* adhesin	*Nicotiana*	Ma *et al.* (1995).
IgM (2)	NP (4-hydroxy 3-nitro-phenylacetyl) (Hapten)	*Nicotiana*	Düring *et al.* (1990).
IgG(k) (2)	Human IgG	*Medicago sativa* (alfalfa)	Khoudi *et al.* (1999)
IgG1 (2)	Herpes simples virus	*Glycine max* (soybean)	Zeitlin *et al.* (1998)
BiscFv	TMV (two separate epitopes)	*Nicotiana*	Fischer *et al.* (1999b)

Table 3. *Some examples of pharmaceutical proteins produced in the tobacco plant system [Table modified from Gruber and Theisen, 2000]*

Protein	Potential Use	Expression level*	Glycosylation	Reference
Human albumin	Shock treatment, burns, co-adjuvant	0.02 %	no	Sijmons *et al.*(1990)
Human haemoglobin	Blood substitute	0.05 %	no	Dieryck *et al* .(1997)
Protein C	Anti-coagulant	0.002 %	no	Cramer *et al.* (1996)
γ-interferon	Phagocyte activator	1 %	yes	Grill L.K. (1997)
Cytokine CM-CSF	Leukopoiesis in bone marrow transplants	NR	yes	Ganz *et al.* (1996)
Epidermal growth factor	Mitogen	0.001 %	yes	Higo *et al.*(1993)
Human erythropoietin	Mitogen, blood cells	0.0026 %	yes	Matsumoto *et al.* (1995)
NP1 defensin	Antibiotic	NR	no	Grill L.K. (1997)
α-galactosidase	Fabry disease	12.1 mg/kg tissue	yes	Grill L.K. (1997)
Glucocere-brosidase	Gaucher disease	1 – 10 %	yes	Cramer *et al.* (1996)
Glutamic acid decarboxylase	Diabetes	0.4 %	no	Ma *et al.*(1997)
Human collagen	Multiple uses	100 mg/kg tissue	yes	Ruggiero *et al.*(2000)
Hepatitis B (surface antigen)	Vaccine	0.01 %	yes	Mason *et al.*(1992)
Enterotoxin B (*E. coli*)	Vaccine (Traveler's diarrhea)	0.001 %	no	Haq *et al.*(1995)
Cholera toxin	Vaccine	NR	no	Hein *et al.*(1996)
Malaria antigen	Vaccine	NR	no	Turpen *et al.*(1995)

* % of extracted proteins -- NR: not reported

ANTIBODIES MADE IN TRANSGENIC PLANTS: PLANTIBODIES

Although antibodies were first expressed in plants in the mid-1980's, by two German graduate students (unpublished theses of During and Steiger; Düring *et al.*, 1990), the first report was published in 1989 (Hiatt *et al.*, 1989). Since then a diverse group of "plantibody" types and forms have been prepared (Table 2). Initially, foreign antibody genes were introduced into plant cells by nonpathogenic strains of the natural plant pathogen *Agrobacterium tumefaciens* (Horsch *et al.*, 1985) and regeneration in tissue culture resulted in the recovery of stable transgenic plants. Although this initial work required crossing of separate plants expressing each chain to generate multichain proteins, more recent studies have shown that multiple chains can be introduced simultaneously. This has been accomplished either by co-cultivation with two *Agrobacterium* strains (de Neve *et al.*, 1993), or bombardment with particles coated with two or more different plasmids (Stoger *et al.*, Wycoff *et al.*, unpublished data), greatly reducing the time to final assembled plantibody.

PLANTS EFFICIENTLY PRODUCE SIGA: A NOVEL ANTIBODY ISOTYPE

This laboratory has focused on the production of secretory IgA (SIgA) plantibodies (Ma *et al.*, 1995, Fig. 1). At the present time, plants offer the only large-scale, commercially viable system for production of this unique form of antibody. SIgA is the most abundant antibody class produced by the body (>60% of total immunoglobulin). SIgA is secreted onto mucosal surfaces to provide local protection from toxins and pathogens. SIgA is comprised of four different protein chains: heavy and light immunoglobulin chains that form the antigen-binding hypervariable region, J chain that dimerizes two IgA molecules (SIgA has four antigen-binding sites) and secretory component that is derived from the mucosal epithelial cells. Dimeric IgA containing J chain derived from submucosal B cells binds to the epithelial cell polyimmunoglobulin receptor (PIG^R). Binding triggers transcytosis to the mucosal surface where a protease releases a portion of the PIG^R called secretory component, conveniently used to bind the SIgA. The secretory component protects the dimeric IgA from proteases and denaturation on the mucosal surface. Previously, it was not possible to obtain therapeutic quantities of this class of immunoglobulin. The recent availability of large amounts of secretory IgA plantibodies opens up a number of novel therapeutic opportunities for disorders of the mucosal immune system. These include therapies for intestinal pathogens such as hepatitis viruses,

Helicobacter pylori, and *enterotoxigenic E. coli*, cholera, etc., respiratory pathogens such as rhinovirus respiratory syncytial virus (RSV) and influenza, and genitourinary sexually transmitted diseases (e.g. HSV) and contraception.

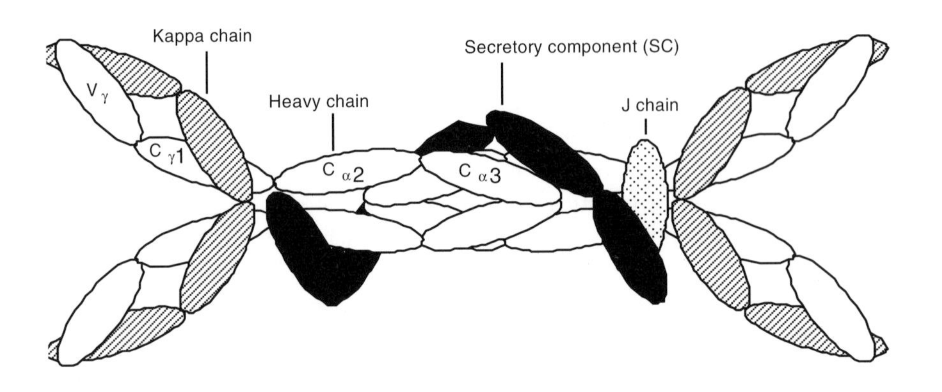

Figure 1. *The Structure of Secretory IgA.*

Table 4. *Plant produced human therapeutics (ca.2000)*

Protein	Target	Organization	Stage
SIgA	Caries	PLANET Biotech	Phase II
IgG	GI cancer	NeoRx/Monsanto	Phase I/II (cancelled due to unexpected gastrointestinal toxicity) (unpublished)
E. coli LT	*E. coli* diarrhea	Boyce Thompson	Phase I/II

To date, three human immunotherapeutic products produced in plants have entered the clinic. These products, listed in Table 4, include two antibodies and an oral vaccine. Other plantibodies in preclinical studies include anti-IgE, C. difficile, and anti-RSV. Clinical studies of the anti-EPCAM plantibody (co-developed by NeoRx and Monsanto) were discontinued due to significant gastrointestinal side-effects (bleeding) unrelated to its production in plants. The anti-S. mutans antibody is currently in phase II trials (see below).

CLINICAL STUDIES OF CARORX™--AN ANTI-*STREPTOCOCCUS MUTANS* SIGA TO PREVENT DENTAL CARIES

The most clinically advanced SIgA plantibody, called **CaroRx™**, recognizes and inhibits the binding of the major oral pathogen, *Streptococcus mutans,* to teeth. In preliminary work, a series of *in vivo* passive immunization experiments was carried out in 84 human subjects using murine anti-*S. mutans* antibodies (Ma *et al.,* 1987, 1989, 1990; Lehner *et al.,* 1975, 1985). Topical application of anti-*S. mutans* antigen SA I/II MAbs prevented colonization of both artificially implanted exogenous strains of *S. mutans*, as well as natural recolonization by indigenous *S. mutans.* In these studies the pathogenic *S. mutans* was hypothesized to be replaced by endogenous flora, i.e. “normal flora” occupy the niche of pathogenic flora.

The presence of the complement-activating and phagocyte-binding sites on the Fc fragment of the MAb was not essential for activity, because the $F(ab')_2$ portion of the MAb was as protective as the intact IgG; however, the Fab fragment failed to prevent recolonization of *S. mutans.* Prevention of recolonization was specifically restricted to *S. mutans*, as the proportion of other organisms, such as *S. sanguis,* did not change significantly. The surprising feature of these experiments was that protection from re-colonization by *S. mutans* lasted up to 2 years (J. Ma, *pers. comm.*), although MAb was applied for only 3 weeks and functional MAb was detected on the teeth for only 3 days following the final application of MAb. All studies indicated that this form of immunotherapy appears to be safe and well tolerated. The long-term protection could therefore not be accounted for by a persistence of MAb on the teeth, but may be due to a shift in the microbial balance in which other bacteria occupy the ecological niche vacated by *S. mutans*, resulting in resistance to recolonization by *S. mutans.*

The antigen-binding V regions of the best murine Mab identified by Ma and Lehner, Guy’s 13, have been used to create an SIgA plantibody produced in tobacco designated **CaroRx™** (details in Ma *et al.,* 1995, 1998). Levels of production of **CaroRx™** in tobacco are up to 0.5 mg/gram fresh weight. Future plans call for production of **CaroRx™** in corn and other cereal grains.

CaroRx™ has been produced and purified from tobacco under GMP conditions for clinical testing in the UK and USA. **CaroRx™** was engineered with an additional IgG CH2 domain to facilitate purification of the antibody by protein G affinity chromatography. A Poros™ Protein G affinity purification was used to obtain >95% pure **CaroRx™** from green plant tissue.

Clinical evaluation of **CaroRx™** in a pilot Phase II trial has been completed at Guy's Hospital, London, UK (Ma *et al.*, 1998). In this trial a functional comparison was made between **CaroRx™** and the parent IgG monoclonal antibody Guy's 13. BIACORE analysis revealed that the affinity of the antibodies for purified *Streptococcus mutans* SA I/II was similar (KD=0.5-1.3 x 10^{-9} M); however, **CaroRx™** had 4-fold higher avidity (functional affinity), a not unexpected result given the tetravalent binding of the SIgA.

Using an experimental design similar to that used to demonstrate activity of the parent Mab, **CaroRx™** gave specific protection against colonization by oral streptococci for over four months (details in Ma *et al.*, 1998). In addition to this therapeutic endpoint, pharmacokinetics studies showed that in the human oral cavity, **CaroRx™** survived for >3 days *versus* 1 day for the IgG antibody and multiple serum antibody samples were negative for human anti-mouse (HAMA) or anti-rabbit antibodies. There was no evidence of local or systemic toxicity of the topically applied plantibody.

These initial clinical studies demonstrate that topically applied anti-*S. mutans* SIgA plantibody (**CaroRx™**) is safe (no antibody response [Human anti-mouse antibody—HAMA], no local or systemic toxicity) and prevents colonization by *S. mutans,* the major cause of human dental caries (Ma *et al..,* 1998). Planet Biotechnology, Inc. has submitted an IND (investigational new drug application) to the US FDA and Phase I/II confirmatory clinical trials were initiated at the School of Dentistry at the University of California in San Francisco in 1998. These studies had an overall design similar to published studies (Ma *et al.,* 1998) comparing 2 treatments (1 week), 4 treatments (2 weeks) and 6 treatments (3 weeks). This study was completed in the Spring of 2000. The full results of this trial will be published elsewhere (Weintraub *et al.,* in prep.). However, initial analysis indicates that the full three weeks of antibody treatment (six topical applications) reduced the number of *S. mutans* to levels that are considered to be clinically significant. Based on these studies, further clinical trials are planned.

Outstanding issues regarding transgenic plant expression of immunotherapeutic proteins

Production costs

Cost benefits of plant production are enormous. For example, Kusnadi *et al.* (1997) calculated the cost of producing a recombinant protein in various agricultural crops. The cost estimate was based on the commodity price of the crop, the fraction of total protein in the crop, and the not unreasonable assumption that the recombinant protein accumulated to 10% of the total plant protein. Although crops with more protein content (*eg.* soybeans-40% versus potatoes-2%) are more cost effective, these costs are 10- to 50-fold less than protein produced at high level in *E. coli* (*i.e.* @20% total protein) (Petridis *et al.,* 1995). Depending upon the use of the protein and the requirements for purification for *in vivo* pharmaceutical use, purification costs will obviously augment final product costs. However, at the hundred kilogram to metric ton level plant produced proteins will provide obvious savings. Our production engineers calculated capitalization costs of plant bioreactors to be 10-20% of steel tank fermentation (unpublished).

Glycosylation of transgenic plant proteins. Plant antibody glycosylation

Cabanes-Macheteau *et al.* (1999) reported the first detailed analysis of the glycosylation of a functional mammalian glycoprotein expressed in a transgenic plant. The structures of the N-linked glycans attached to the heavy chains of the monoclonal antibody Guy's 13 produced in transgenic tobacco plants (plantibody Guy's 13) were identified and compared to those found in the corresponding IgG1 of murine origin. As in mouse antibodies, both N-glycosylation sites located on the heavy chain of the plantibody Guy's 13 are N-glycosylated. However, the number of Guy's 13 glycoforms is higher in the plant than in the mammalian expressed antibodies. In addition to high-mannose-type N-glycans, 60% of the oligosaccharides N-linked to the plantibody have beta(1, 2)-xylose and alpha(1,3)-fucose residues linked to the core Man3GlcNAc2.

These oligosaccharides linkages, not found on mammalian N-linked glycans, are potentially immunogenic and raise the possibility that plantibodies containing N-linked glycans may have a limited scope as parenterals or even when applied topically or orally, particularly in patients with severe food allergies. Food allergans bearing beta(1, 2)-xylose and alpha(1,3)-fucose have been linked to specific IgE in serum and to biological activity, i.e. histamine release, in allergic patients (Fotisch *et al.*, 1999 and Garcia-Casado *et al.*, 1996). In contrast, the mere presence of serum IgE against cross-reactive carbohydrate determinants, which include the beta(1,2)-xylose and alpha(1,3)-fucose linkages to the core Man3GlcNAc2 of plant glycans, has been shown to be a poor predicter of clinical allergy (van Ree and Aalberse, 1999; Mari *et al.*, 1999; van der Veen *et al.*, 1997). These results preclude generalities about the potential toxicity of plantibody glycans in humans.

When the mouse plantibody (mouse amino acid sequence and plant glycans) was used to immunize mice there was no, or only a minimally detectable, serum immune response (Chargelegue, personal communication). Thus, in the mouse at least, a "self" primary protein structure, decorated with plant N-linked glycans, can be non-immunogenic. This plantibody has been applied topically in the mouth of humans with no detection of human anti-mouse antibodies (Ma *et al.*, 1998; Planet Biotechnology Phase II results).

Humanization of plantibody primary structures may go a long way to obviate the immunogenic potential of plant glycans and further genetic engineering may, if necessary, alter the glycan structures themselves. Most dramatically, aglycosyl antibodies can be created by altering the peptide recognition sequence for N-linked glycosylation (asn-X-Ser/Thr). High-mannose type glycosylation, which does not contain the core-linked xylose and fucose residues, may be favored by the addition of a C-terminal KDEL sequence (Wandelt *et al.*, 1992, Bednarek *et al.*, 1992) and the subsequent targeting of plantibodies to the proximal endoplasmic reticulum. Alternatively, the specific fucosyl (Leiter *et al.*, 1999) or xylosyl transferases, which operate in the trans-golgi, may be targeted for gene silencing.

Gene silencing: a potential problem of plant expression

Many labs have noted the common occurrence of transgene silencing in green plants. For example, de Neve *et al.* (1999) assessed the stability of antibody and Fab expression in five different homozygous transgenic *Arabidopsis* lines. Each of these lines showed silencing of the transgenes that encoded the antibody polypeptides, leading to instability of antibody production. However, each line had a different and specific instability profile. The characteristic variation in the level of antibody accumulation in each line as a function of developmental stage indicated that the T-DNA integration pattern played a role in triggering silencing, and also that the history and the integration position of simple transgene loci can influence the susceptibility to epigenetic silencing. In different lines with low antibody accumulation levels, methylation was found either in the promoter alone, or in both the promoter and the transcribed region; in the transcribed region only, or in the transcribed region and downstream sequences. Epigenetic effects resulted in different transgene expression profiles in each of the five *Arabidopsis* lines analyzed. In our own experience using tobacco, gene silencing has been a difficult problem. With the first generation plantibody constructs described above using the 35S cauliflower mosaic virus promoter introduced via *Agrobacterium* transformation, gene silencing has been less of a problem than with recent constructs using much stronger promoters and biolistic transformation. A relationship between transcript abundance, sequence redundancy within the constructs and other factors has been observed. Work is underway with novel constructs and means of delivery to alleviate this important problem. Knock out of the RNA-dependent RNA polymerase gene may be another means to alleviate this problem.

Purification and process development of plantibodies

The potential for cost reduction when using genetically engineered green plants as bioreactors, instead of conventional pharmaceutical factories engineered with concrete and steel, is a powerful argument for the use of transgenic plants to produce recombinant proteins. The commercial scale production of proteins from transgenic plants generally requires one to envisage the growth and processing of tons of biomass to achieve the economy of scale which would fully exploit the inputs of sunlight, soil, water and fertilizer. The processing of large amounts of biomass also anticipates large numbers of patients or consumers, large amounts of purified protein needed for each patient or consumer, lower than anticipated levels of expression, and losses during processing.

The realities of commercial-scale production benefit from a demand for simplicity early on in the development of a large-scale process. Bench-scale or laboratory-scale procedures often employ the pampering conditions that are necessary for a proof of concept but which are too complicated and expensive for large-scale efforts. For instance, because of issues regarding toxicity and expense, it is preferred not to use protease inhibitors beyond the bench scale. Reagents such as ammonium sulfate are frowned upon because of the disposal-associated costs, as are organic solvents because of their toxicity and flammability. Efforts to avoid the purchase and maintenance of centrifuges will eventually be gratefully acknowledged by maintenance personnel. In addition, and more fundamentally, it is important to realize that a commercial process will not be executed by rocket scientists, each with extensive post-doctoral training, but by more ordinary, and while no less dedicated, almost certainly less well trained and educated people. Thus simplicity in the number, as well as the type, of components in a process is paramount to reduce not only costs but also errors.

Minimalist approaches to large-scale process development will often be rewarded, even though they may appear simplistic. We found that grinding transgenic tobacco in water alone gave up to 70% of the expected immunoglobulin compared to tobacco ground in a buffer, poised at a specified pH, containing 6 components in addition to water; each component had been chosen for a particular, biochemically sound, reason. Yet this was evidence which suggested that perhaps not all of the components were vitally important and we now use a two-component buffer at a specified pH.

Commercial-scale production of recombinant proteins from plants will also benefit from the technology and equipment commonly used in the food and beverage industry. Grain mills and cole-slaw slicers will doubtless be useful off-the-shelf machinery for the initial processing of seeds and leafy tissue. The beverage industry has always been concerned with the clarification of juices and their phenolic content (van Sumere *et al.*, 1975) and is a source of knowledge and equipment, both new and used. The chef's trick of delaying the browning of cut fruit by the addition of lemon acidulated water (Bombauer and Becker, 1975) has led us to formulate buffers to inhibit the "tanning" of proteins in tobacco extracts.

Two constituents of stem and leaf extracts which require special consideration are membranes and cell walls. The green color of stem and leaf extracts indicates the presence of chlorophyll and the suspension of thylakoid membranes. These and other membranes must be removed to allow filterability at or below a pore size of 0.45 μm. This level of filterability helps to ensure, but does not guarantee, the good behavior of the extract during subsequent chromatography and serves to remove bacterial sources of contamination. The cellulose-containing debris, which forms a large part of the insoluble portion of extracts, can be used as an endogenous "filter aid" during the initial clarification to remove these membranous components.

Phenolics are a major concern when extracting most stems and leaves and one's efforts are rewarded by their early removal. They may interact with proteins, and other extract components, via hydrophobic interactions, salt bridges, hydrogen bonding and by additional reactions to nucleophilic centers (Gegenheimer, 1990; van Sumere *et al.*, 1975; Loomis, 1974). These interactions can dramatically and irreversibly alter the properties of proteins. Fortunately the majority of released phenolics are generally small in size, as well as water soluble, and may be removed by tangential-flow ultrafiltration/diafiltration, which also serves to concentrate the considerably larger proteins of interest. In addition, other incompatible, water soluble secondary metabolites, such as neonicotine (anabasine) and nicotine from tobacco, may also be removed by ultrafiltration/diafiltration.

After such clarification and concentration steps, recombinant proteins can be assumed to behave independently and with regard to the peculiarities of their own biochemistry. In other words, they are now ready for chromatography. Residual phenolics may still dictate the degree of clean-up necessary before chromatography on an expensive Protein A or Protein G affinity column is allowed; these are economic considerations common to any purification process.

One additional note of caution: since the processed plant or seed was probably grown in soil, or some similarly unclean stratum, one may anticipate the presence of a diverse bioburden. Contamination of product with endotoxins and mycotoxins can be minimized by rapid processing and early filtration but process development must also always recognize the necessity to eliminate compounds as well as the necessity to purify, concentrate and stabilize a product of interest.

SUMMARY

Antibodies represent the largest class of biopharmaceutical molecules under development. Plants offer a cost-effective bioreactor to produce antibodies of diverse types. Recent studies demonstrate that secretory IgA, the predominant antibody isotype of the mucosal immune system, can be made in large quantities in plants. **CaroRx**™, the lead SIgA antibody being developed by Planet Biotechnology, Inc. has demonstrated activity in pilot Phase II trials versus *S. mutans,* the major pathogen contributing to development of dental caries. Numerous other SIgA plantibodies are in pre-clinical development.

REFERENCES

Arakawa T, Chong DK and Langridge WH. 1998. Efficacy of a food plant-based oral cholera toxin B subunit vaccine. *Nat. Biotechnol.* 16(3):292-297.

Artsaenko O, Peisker M, zur Nieden U, Fiedler U, Weiler EW, Müntz K and Conrad U. 1995. Expression of a single-chain Fv antibody against abscisic acid creates a wilty phenotype in transgenic tobacco. *Plant J.*, 8:745-750.

Austin S, Bingham ET, Koegel RG, Mathews DE, Shahan MN, Straub RJ and Burgess RR. 1994. An overview of a feasibility study for the production of industrial enzymes in transgenic alfalfa. *Ann. N.Y. Acad. Sci.*, 721: 235-244.

Baum TJ, Hiatt A, Parrott WA, Pratt LH and Hussey RS. 1996. Expression in tobacco of a functional monoclonal antibody specific to stylet secretions of the root-knot nematode. *Molecular Plant Microbe Interactions*, 9:382-387.

Bednarek SY and Raikhel NV. 1992. Intracellular trafficking of secretory proteins. *Plant Mol. Biol.*,20: 133-150.

Benvenuto E, Ordas RJ, Tavazza R., Ancora G, Biocca S, Cattaneo A and Galeffi P. 1991. "Phytoantibodies"; A general vector for the expression of immunoglobulin domains in transgenic plants. *Plant Mol. Biol.*, 17:865-874.

Bombauer IvS and Becker MR. 1975. In, Joy of Cooking, 38th edition, Bobbs-Merril Co., Inc. (Indianapolis):p 520.

Borrebaeck CAK. (ed.) 1995. Antibody Engineering. Oxford University Press. New York, Oxford.

Bruyns A-M, de Jaeger G, De Neve M, De Wilde C, Van Montagu M and Depicker A. 1996. Bacterial and plant-produced scFv proteins have similar antigen-binding properties. *FEBS Lett.*, 386:5-10.

Cabanes-Macheteau M, Fitchette-Laine AC, Loutelier-Bourhis C, Lange C, Vine ND, Ma JK, Lerouge P and Faye L. 1999. N-Glycosylation of a mouse IgG expressed in transgenic tobacco plants. *Glycobiology*, 9:365-72.

Casadevall A and Scharff MD. 1994. "Serum therapy" revisited: Animal models of infection and the development of passive antibody therapy. *Antimicrob. Agents Chemother.*, 38:1695-1702.

Casadevall A, *et al.* 1997. Antibody-based therapies for infectious diseases: Renaissance for an abandoned arsenal? *Bull. Inst. Pasteur*, 95:247, 1997.

Casadevall A and Scharff, MD. 1995. Return to the past: the case for antibody-based therapies in infectious diseases. *Clin. Infect. Dis.,* 21: 150-161.

Conrad U and Fiedler U. 1998. Compartment-specific accumulation of recombinant immunoglobulins in plant cells: an essential tool for antibody production and immunomodulation of physiological functions and pathogen activity. *Plant Mol. Biol.*, 38:101-9.

Cramer CL, Weissenborn DL, Oishi KK, Grabau EA, Bennett S, Ponce E, Grabowski GA and Radin DN. 1996. In "Engineering plants for commercial products and applications", G.B. Collins and R.J. Shepherd, Eds., New York Academy of Sciences, New York, p. 62.

de Haard H, Henderikx P and Hoogenboom HRP. 1998. Creating and engineering human antibodies for immunotherapy. *Adv. Drug Delivery Rev.,* 31:5-31.

de Neve M, De Buck S, De Wilde C, Van Houdt H, Strobbe I, Jacobs A, Van Montagu and Depicker A. 1999. Gene silencing results in instability of antibody production in transgenic plants. *Mol. Gen. Genet.,* 260:582-592.

de Neve M, De Loose M, Jacobs A, Van Houdt H, Kaluza B, Weidle U and Depicker A. 1993. Assembly of an antibody and its derived antibody fragment in *Nicotiana* and *Arabidopsis*. *Transgenic Research*, 2:227-237.

de Wilde C, de Neve M, de Rycke R, Bruyns A.M, de Jaeger G, van Montagu M, Depicker A and Engler G. 1996. Intact antigen-binding MAK33 antibody and Fab fragment accumulate in intercellular spaces of *Arabidopsis thaliana*. *Plant Sci.*, 114:233-241.

de Wilde C, de Rycke R, TB, de Neve M, van Montagu M, Engler G and Depicker A. 1998. Accumulation pattern of IgG antibodies and Fab fragments in transgenic *Arabidopsis thaliana* plants. *Plant Cell Physiology*, 39:639-646.

Dieryck W, Pagnier J, Poyart C, Marden MC, Gruber V, Bournat P, Baudino S and Mero B. 1997. Human haemoglobin from transgenic tobacco. *Nature* 386, 29-30.

Düring K, Hippe S, Kreuzaler F and Schell J. 1990. Synthesis and self-assembly of a functional monoclonal antibody in transgenic *Nicotiana tabacum*. *Plant Mol. Biol.*, 15:281-293.

Fecker LF, Kaufmann A. Commandeur U., Commandeur J., Koenig R. and Burgermeister W. 1996. Expression of single-chain antibody fragments (scFv) specific for beet necrotic yellow vein virus coat protein or 25 kDa protein in Escherichia coli and *Nicotiana benthamiana*. *Plant Mol. Biol.*, 32:979-986.

Fecker LF, Koenig R and Obermeier C. 1997. Nicotiana benthamiana plants expressing beet necrotic yellow vein virus (BNYVV) coat protein-specific scFv are partially protected against the establishment of the virus in the early stages of infection and its pathogenic effects in the late stages of infection. *Archives of Virology*, 142:1857-1863.

Fiedler U and Conrad U. 1995. High-level production and long-term storage of engineered antibodies in transgenic tobacco seeds. *Bio/Technol.*, 13:1090-1093.

Fiedler U, Phillips J, Artsaenko O and Conrad U. 1997. Optimization of scFv antibody production in transgenic plants. *Immunotechnology*, 3:205-216.

Firek S, Draper J, Owen MRL, Gandecha A, Cockburn B, and Whitelam GC. 1993. Secretion of a functional single-chain Fv protein in transgenic tobacco plants and cell suspension cultures. *Plant Mol. Biol.*, 23:861-870.

Fischer R, Liao Y-C and Drossard J. 1999. Affinity purification of a TMV-specific recombinant full-size antibody from a transgenic tobacco suspension culture. *J. Immunol. Methods*, 226:1-10.

Fischer R, Liao Y-C, Hoffman K, Schilberg S and Emans N. 1999a. Molecular farming of recombinant antibodies in plants. *Biol. Chem.*, 380:825-839.

Fischer R, Schumann D, Zimmerman S, Drossard J, Sack M and Schillberg S. 1999b. Expression and characterization of bispecific single chain Fv fragments produced in transgenic plants. *Eur. J. Biochem.*, 262:810-816.

Fotisch K, Altmann F, Haustein D and Vieths S. 1999. Involvement of carbohydrate epitopes in the IgE response of celery-allergic patients. *Int. Arch. Allergy Immunol.*, 120(1):30-42.

Fraley RT, Rogers SG, Horsch RB, Sanders PR, Flick JS, Adams SP, Bittner ML, Brand LA, Fink CL, Fry JS, Galluppi GR, Goldberg SB and Hoffmann NL, Woo S.C. 1983. Expression of bacterial genes in plant cells. *Proc. Natl. Acad. Sci. USA* 80: 4803-4807.

Franconi R, Roggero P, Pirazzi P, Arias FJ, Desiderio A, Bitti O, Pashkoulov D, Mattei B, Bracci L, Masenga V, Milne RG and Benvenuto E. 1999. Functional expression in bacteria and plants of an scFv antibody fragment against tospoviruses. *Immunotechnology*, 4:189-201.

Ganz PR, Dudani AK, Tackaberry ES, Sardana R, Sauder C, Cheng X and Altosaar I. 1996. In "Transgenic plants: a production system for industrial and pharmaceutical proteins". M.R.L. Owen and J. Pen, Eds., Wiley, Chichester, p 281.

Garcia-Casado G, Sanchez-Monge R, Chrispeels MJ, Armentia A, Salcedo G and Gomez L. 1996. Role of complex asparagine-linked glycans in the allergenicity of plant glycoproteins. *Glycobiology*. 4:471-477.

Gavilondo JV and Larrick JW. 2000. Antibody engineering at the Millenium. BioTechniques.

Gegenheimer P. 1990. Preparation of Extracts from Plants. Chap 14. In, *Methods Enzym.*, 182:174-193.

Grill LK. 1997. In "IBC's 3rd Annual International Symposium on Producing the Next Generation of Therapeutics: Exploiting Transgenic Technologies", West Palm Beach, 5-6 Feb.

Gruber V and Theisen M. 2000. Genetically Modified Crops as a Source for Pharmaceuticals. *Ann. Reports in Medicinal Chemistry* 35, chapter 31, pp 357-363.

Haq TA, Mason HS, Clements JD and Arntzen CJ. 1995. Oral immunization with a recombinant bacterial antigen produced in transgenic plants. *Science,* 268: 714-716.

Hein M, Tang Y, McLeod DA, Janda KD and Hiatt AC. 1991. Evaluation of Immunoglobulins from plant cells. *Biotechnology Progress*, 7:455-461.

Hein MB, Yeo TC, Wang F and Sturtevant A. 1996. In "Engineering plants for commercial products and applications", G.B. Collins and R.J. Shepherd, Eds., New York Academy of Sciences, New York, p 50.

Hiatt A. 1990. Antibodies produced in plants. *Nature*, 344:469-470.

Hiatt A. 1991. Monoclonal antibodies, hybridoma technology and heterologous production systems. *Current Opinion in Immunology*, 3:229-232.

Hiatt A, Cafferkey R and Bowdish K. 1989. Production of antibodies in transgenic plants. *Nature*, 342:76-78.

Hiatt A and Ma J. 1993. Characterization and applications of antibodies produced in plants. *International Review of Immunology*, 10:139-152.

Hiatt A and Ma JK-C. 1992. Monoclonal antibody engineering in plants. *FEBS Lett.*, 307:71-75.

Hiatt A, Tang Y, Weiser W and Hein MB. 1992. Assembly of antibodies and mutagenized variants in transgenic plants and plant cell cultures. *Genetic Engineering*, 14:49-64.

Higo K, Saito Y and Higo H. 1993. Expression of a chemically synthesized gene for human epidermal growth factor under the control of cauliflower mosaic virus 35S promoter in transgenic tobacco. *Biosci. Biotechnol. Biochem.* 57, 1477-1481.

Hood EE and Jilka JM. 1999. Plant-based production of xenogenic proteins. *Curr. Opin. Biotechnol.*,10:382-386.

Horsch RB, Fry JE, Hoffmann NL, Eichholtz D, Rogers SG and Fraley RT. 1985. A simple and general method for transferring genes into plants. *Science*, 227:1229-1231.

Horvath H, Huang J, Wong O, Kohl E, Okita T, Kannangara LG and von Wettstein D. 2000. The production of recombinant proteins in transgenic barley grains. *Proc. Natl. Acad. Sci. USA* 97:1914-1919.

Kapusta J, Modelska A, Figlerowicz M, Pniewski T, Letellier M, Lisowa O, Yusibov V, Koprowski H, Plucienniczak A and Legocki AB. 1999. A plant-derived edible vaccine against hepatitis B virus. *FASEB J.*, 13:1796-1799.

Khoudi H, Laberge S, Ferullo JM, Bazin R, Darveau A, Castonguay Y, Allard G, Lemieux R and Vezina LP. 1999. Production of a diagnostic monoclonal antibody in perennial alfalfa plants. *Biotechnol. Bioeng.*, 20:135-143.

Kusnadi Ann R, Nikolov Z and Howard John A. 1997. Production of Recombinant Proteins in Transgenic Plants: Practical Considerations. *Biotechnol. Bioengineer.,* 56:473-484.

Kusnadi AR, Evangelista RL, Hood EE, Howard JA and Nikolov ZL. 1998. Processing of transgenic corn seed and its effect on the recovery of recombinant beta-glucuronidase. *Biotechnol. Bioeng.* Oct 5;60(1):44-52.

Kusnadi AR, Hood EE, Witcher DR, Howard JA and Nikolov ZL. 1998. Production and purification of two recombinant proteins from transgenic corn. : *Biotechnol. Prog.* Jan-Feb;14(1):149-155.

Kusnadi, A. *et al.* 1997. Recovery of recombinant beta-glucuronidase from transgenic corn. In: L.E. Erickson (ed.), The proceedings of the 26th annual Biochemical Engineering Symposiums (pp. 143-148), Kansas State University, Manhattan, KS.

Larrick JW, Yu L, Chen J, Jaiswal S and Wycoff K. 1998. Production of antibodies in transgenic plants. *Research Immunology*, 149:603-608.

Lehner T, Caldwell J and Smith R. 1985. Local passive immunization by monoclonal antibodies against streptococcal antigen I/II in the prevention of dental caries. *Infection and Immunity*, 50: 796.

Lehner T, Challaombe SJ and Caldwell J. 1975. Immunological and bacteriological basis for vaccination against dental caries in rhesus monkeys. *Nature*, 254:517.

Leiter H, Mucha J, Staudacher E, Grimm R, Glossl J, Altmann F. 1999. Purification, cDNA cloning, and expression of GDP-L-Fuc:Asn-linked GlcNAcalpha1,3-fucosyltransferase from mung beans. *J. Biol. Chem.* Jul 30;274(31):21830-21839.

Loomis W.D. 1974. Overcoming problems of Phenolics and Quinones in the Isolation of Plant

Enzymes and Organelles. Chap. 54. In: *Methods Enzym.*, 31 (PtA):528-544.

Ma JK-C, Hunjan M, Smith R and Lehner, T. 1989. Specificity of monoclonal antibodies in local passive immunization against *Streptococcus mutans*. *Clin. Exp. Immunol.*, 77:331-337.

Ma JK-C and Lehner T. 1990. Prevention of colonization of *Streptococcus mutans* by topical application of monoclonal antibodies in human subjects. *Archs. Oral Biol.*, 35:115S-122S.

Ma JK-C, Smith R and Lehner, T. 1987. Use of monoclonal antibodies in local passive immunization to prevent colonization of human teeth by *Streptococcus mutans*. *Infection and Immunity*, 55:1274-1278.

Ma SW, Zhao DL, Yin ZQ, Mukherjee R, Singh B, Qin HY, Stiller CR and Jevnikar AM. 1997. Transgenic plants expressing autoantigens fed to mice to induce oral immune tolerance. *Nat. Med.*3:793-796.

Ma J, Hikmat B, Wycoff K, Vine N, Chargelegue D, Yu L, Hein M and Lehner T. 1998. Characterization of a recombinant plant monoclonal secretory antibody and preventive immunotherapy in humans. *Nat. Med.*, 4:601-606.

Ma J.K, Hiatt A, Hein M, Vine ND, Wang F, Stabila P, van Dolleweerd C, Mostov K and Lehner T. 1995. Generation and assembly of secretory antibodies in plants. *Science*, 268:716-719.

Ma JK-C, Lehner T, Stabilia P, Fux C.I and Hiatt A. 1994. Assembly of monoclonal antibodies with IgG1 and IgA heavy chain domains in transgenic tobacco plants. *Eur. J. Immunol.*, 24:131-138.

Ma S, Zhao D, Yin A, Mukherjee R, Singh B, Qin H, Stiller CR and Jevnikar A.M. 1997. *Nat. Med.* 3, 793-796.

Mari A, Iacovacci P, Afferni C, Barletta B, Tinghino R, Di Felice G and Pini C. 1999. Specific IgE to cross-reactive carbohydrate determinants strongly affects the in vitro diagnosis of allergic diseases. *J. Allergy Clin. Immunol.*, 103:1005-1011.

Mason H.S, Ball JM, Shi JJ, Jiang X, Estes MK and Arntzen CJ. 1996. Expression of Norwalk virus capsid protein in transgenic tobacco and its oral immunogenicity in mice. *Proc. Natl. Acad. Sci. USA* 93: 5335-5340.

Mason HS, Lam DM and Arntzen CJ. 1992. Expression of hepatitis B surface antigen in transgenic plants. *Proc. Natl. Acad. Sci. USA,* 89: 11745-11749.

Mason HS, Haq TA, Clements JD and Arntzen CJ. 1998. Edible vaccine protects mice against Escherichia coli heat-labile enterotoxin (LT): potatoes expressing a synthetic LT-B gene. *Vaccine* 16:1336-1343.

Matsumoto S, Ikura K, Ueda M and Sasaki R. 1995. Characterization of a human glycoprotein (erythropoietin) produced in cultured tobacco cells. *Plant Mol. Biol.* 27, 1163-1172.

McCormick AA, Kumagai MH, Hanley K, Turpen TH, Hakim I, Grill LK, Tuse D, Levy S and Levy R. 1999. Rapid production of specific vaccines for lymphoma by expression of the tumor-derived single-chain Fv epitopes in tobacco plants. *Proc. Natl. Acad. Sci. USA*, 96:703-708.

Owen M, Gandecha A, Cockburn B and Whitelam G. 1992. Synthesis of a functional anti-phytochrome single-chain Fv protein in transgenic tobacco. *BioTechnol.*, 10:790-794.

Petridis D, Sapidou E and Calandranis J. 1995. Computer-aided process analysis and economic evaluation for biosynthetic human insulin production – a study case. *Biotechnol. Bioeng.*, 48, 529-541.

Phillips J, Artsaenko O, Fiedler U, Horstmann C, Mock HP, Muntz K and Conrad U. 1997. Seed-specific immunomodulation of abscisic acid activity induces a developmental switch. *EMBO J.*, 16:4489-4496.

Pollock DP, Kutzko JP, Birck-Wilson E, Williams JL, Echelard Y and Meade HM. 1999. Transgenic milk as a method for the production of recombinant antibodies. *J. Immunol.. Methods,* 231:147-157.

Rosso M-N, Schouten A, Roosien J, Borst-Vrenssen T, Hussey RS, Gommers FJ, Bakker J, Schots A and Abad P. 1996. Expression and functional characterization of a single chain FV antibody directed against secretions involved in plant nematode infection process. *Biochem. Biophys. Res. Commun.*, 220:255-263.

Ruggiero F, Exposito J-Y, Bournat P, Gruber V, Perret S, Comte J, Olagnier B, Garrone R and Theisen M. 2000. Triple helix assembly and processing of human collagen produced in transgenic plants. *FEBS Lett.,* 469, 132-136.

Russell DA. 1999. Feasibility of antibody production in plants for human therapeutic use. *Curr. Top. Microbiol. Immunol.*, 236:119-137.

Salmon V, Legrand D, Slomianny MC, El Yazidi I, Spik G, Gruber V, Bournat P, Olagnier B, Mison D, Theisen M and Merot B. 1998. Production of human lactoferrin in transgenic tobacco plants. *Protein Express. Purif.* 13, 127-135.

Sanford JC. 1988. The Biolistic process. *Trends in Biotechnology*, 6:299-302.

Schillberg S, Zimmermann S, Voss A and Fischer R. 1999. Apoplastic and cytosolic expression of full-size antibodies and antibody fragments in *Nicotiana tabacum. Transgenic Research*, 8:255-263.

Schouten A, Roosien J, de Boer JM, Wilmink A, Rosso MN, Bosch D, Stiekema WJ, Gommers FJ, Bakker J and Schots A. 1997. Improving scFv antibody expression levels in the plant cytosol. *FEBS Lett.*, 415:235-241.

Schouten A, Roosien J, van Engelen FA, de Jong GAM, Borst-Vrenssen AWM, Zilverentant JF, Bosch D, Steidema WJ, Gommers FJ, Schots A and Bakker J. 1996. The C-terminal KDEL sequence increases the expression level of a single-chain antibody designed to be targeted to both the cytosol and the secretory pathway in transgenic tobacco. *Plant Mol. Biol.*, 30:781-793.

Sijmons PC, Dekker BMM, Schrammeijer B, Verwoerd TC, van den Elzen PJM and Hoekma A. 1990. Production of correctly processed human serum albumin in transgenic plants. *BioTechnology*, 8, 217-221.

Stoger E, Vaquero C, Torres E, Sack M, Nicholson L, Drossard J, Williams S, Keen D, Perrin Y, Christou P and Fischer R. 2000. Cereal crops as viable production and storage systems for pharmaceutical scFv antibodies, *Plant Mol. Biol.* 42, 583-90.

Tavladoraki P, Benvenuto E, Trinca S, Demartinis D, Cattaneo A and Galeffi P. 1993. Transgenic plants expressing a functional single-chain Fv antibody are specifically protected from virus attack. *Nature*, 366:469-472.

Tavladoraki P, Girotti A, Donini M, Arias FJ, Mancini C, Morea V, Chiaraluce R, Consalvi V and Benvenuto E. 1999. A single-chain antibody fragment is functionally expressed in the cytoplasm of both *Escherichia coli* and transgenic plants. *Eur. J. Biochem.*, 262:617-624.

Turpen TH, Reinl SJ, Charoenvit Y, Hoffman SL, Fallarme V and Grill LK. 1995. Malarial epitopes expressed on the surface of recombinant tobacco mosaic virus. *BioTechnol.* 13, 53-57.

van der Veen MJ, van Ree R, Aalberse RC, Akkerdaas J, Koppelman SJ, Jansen HM and van der Zee JS. 1997. Poor biologic activity of cross-reactive IgE directed to carbohydrate determinants of glycoproteins. *J. Allergy Clin. Immunol.,* 100:327-34.

van Engelen FA, Schouten A, Molthoff JW, Roosien J, Salinas J, Dirkse WG, Schots A, Bakker J, Gommers FJ, Jongsma MA, Bosch D and Steikema WJ. 1994. Coordinate expression of antibody subunit genes yields high levels of functional antibodies in roots of transgenic tobacco. *Plant Mol. Biol.*, 26:1701-1710.

van Ree R and Aalberse RC. 1999.Specific IgE without clinical allergy. *J Allergy Clin. Immunol.*,103:1000-1001.

Van Sumere CF, Albrecht J, Dedonder A, de Pooter H and Pé I. 1975. Plant Proteins and Phenolics. Chap. 8 in The Chemistry and Biochemistry of Plant Proteins (Harborne, J.B. and van Sumere, C.F., eds.), Academic Press (London); p 211-264.

Vaquero C, Sack M, Chandler J, Drossard J, Schuster F, Monecke M, Schillberg S and Fischer R. 1999. Transient expression of a tumor-specific single-chain fragment and a chimeric antibody in tobacco leaves. *Proc. Natl. Acad. Sci. USA*, 96:11128-33.

von Behring E and Kitasato S. 1890. Ueber Zustandekommen der Diphtherie-Immunitat und der Tetanus-Immunitat bei Thieren. S. *Dtsch. Med. Wochenschr.* 16:1113.

Wandelt CI, Khan MRI, Craig S, Schroeder HE, Spencer D and Higgins TJV. 1992. Vicilin with carboxy-terminal KDEL is retained in the endoplasmic reticulum and accumulates to high levels in leaves of transgenic plants. *Plant* Jour. 2(2):181-192.

Whitelam GC, Cockburn W and Owen MRL. 1994. Antibody production in transgenic plants.

Biochemical Society Transactions, 22:940-943.

Zambryski, P. *et al.* 1983. Ti plasmid vector for the introduction of DNA into plant cells without alteration of their normal regeneration capacity. *EMBO J.*, 2: 2143-2150.

Zeitlin L, Olmsted SS, Moench TR, Co MS, Martinell BJ, Paradkar VM, Russell DR, Queen C, Cone RA and Whaley KJ. 1998. A humanized monoclonal antibody produced in transgenic plants for immunoprotection of the vagina against genital herpes. *Nat. Biotechnol.*, 16:1361-1364.

Zimmerman S, Schillberg S and Liao YC. 1998. Intracelluar expression of TMV-specific single-chain Fv fragments leads to improved virus resistance in *Nicotiana tabacum*. *Molecular Breeding*, 4:369-379.

ANIMAL HEALTH

Joseph M. Jilka
Stephen J. Streatfield

ProdiGene, Inc.
101 Gateway Boulevard
College Station, TX 77845 USA

INTRODUCTION

Recent technological advances in transgenic plants expressing recombinant proteins have resulted in the possibility of paradigm shifts in the area of animal healthcare (Hood *et al.*, 1997). By expressing proteins in edible tissue, the possibility of delivering a healthcare product directly by ingestion of the edible tissue is now becoming a reality. In addition to the potential ease of application, the low cost of production and the enormous amounts of product that can be produced in a plant-based system offer the possibility of disease control on a large scale. The types of products possible to be delivered in this manner could include therapeutic proteins, growth promotants, vaccines, monoclonal antibodies and proteins with antimicrobial activity. To date most activity in this area has focused upon the development of plant-based vaccines.

The utilization of transgenic plants for the delivery of animal healthcare has several potential benefits over traditional routes.

- Transgenic plants are usually constructed to express only a small antigenic portion of the pathogen or toxin, eliminating the possibility of infection or innate toxicity and reducing the potential for adverse reactions as in the case of whole virus vaccines.
- Since there are no known human or animal pathogens that are able to infect plants, concerns with viral or prion contamination are eliminated.
- Production in transgenic crops relies on the same established technologies to sow, harvest, store, transport, and process the plant as those commonly used for food crops, making transgenic plants a very economical means of large-scale vaccine production. Large quantities of protein-based therapeutics could be produced very economically.

E.E. Hood and J.A. Howard (eds.), Plants as Factories for Protein Production, 103–118.

- Expression of the recombinant protein comprising the therapeutic or vaccine in the natural protein-storage compartments of plants maximizes stability, minimizes the need for refrigeration and keeps transportation and storage costs low.
- Formulation of multicomponent therapeutics or vaccines is possible by blending the seed of multiple transgenic corn lines into a single vaccine.
- Direct oral administration is possible when immunogens are expressed in commonly consumed feed plants, such as grain, leading to the production of edible vaccines. In the case of vaccines, the immunogen could be produced in a safe, directly edible or easily purified form for edible, oral, or even parenteral vaccines. These vaccines could be used in a stand-alone vaccination strategy, as a booster, or in combination with other vaccines and vaccination routes.

Edible therapeutics and vaccines from plant material could be directly delivered in the feed and could be produced cheaply in large volumes thus avoiding many costs associated with the administration of conventional treatments. The use of transgenic plant material for animal health has, to date, been limited to the field of edible vaccines. Consequently, it is the focus of this review to highlight results obtained in the development of edible vaccines to animal diseases. Vaccines from plants are particularly suitable for stimulation of mucosal immunity, since edible plant products can be delivered orally to reach the gut mucosal tissue and elicit an immune response at mucosal surfaces. Recent advances in technology make it now possible to express vaccine antigens at high levels in plants (Streatfield *et al.*, 2001).

With the advent of improved transformation technology in the past decade, transgenic plants have now been successfully used to express a variety of genes. Numerous genes have been cloned into a variety of transgenic plants including many enzymes that demonstrate the same activity as their authentic counterparts (Hood *et al.*, 1997, Pen *et al.*, 1992, Trudel *et al.*, 1997). In 1999 over 60 million acres of transgenic plants were grown indicating that the system of production of transgenic plants is stable and robust. In particular, many additional genes have been expressed in plants solely for their immunogenic potential, including viral proteins (Lee *et al.*, 2001, Daniell *et al.*, 2001, Gomez *et al.*, 1999, Mason *et al.*, 1996, McGarvey *et al.*, 1995, Thanavala *et al.*, 1995, Modelska *et al.*, 1998, Wigdorovitz *et al.*, 1999) and subunits of bacterial toxins (Arakawa *et al.*, 1997, Arakawa *et al.* 1999, Haq *et al.*, 1995, Mason *et al.,* 1998). Animal and human immunization studies have demonstrated the effectiveness of many of these plant-derived recombinant antigens in stimulating the immune system. The production of antigen-specific antibodies and protection against subsequent toxin or pathogen challenge demonstrate the feasibility of plant derived-antigens for immunological use. Many of these plant-derived antigens induced an immunological response comparable to that of the antigens in the original pathogen in mice (Gomez *et al*, 1998, Mason *et al*, 1998, Modelska *et al.*, 1998, Wigdorovitz *et al.*, 1999) humans (Kapusta *et al.*, 1999), poultry (ProdiGene, unpublished data) and swine (Streatfield *et al.*, 2001). Characterization studies of these engineered immunogens have shown that plants have the ability to express, fold and modify proteins in a manner that is consistent with the authentic source.

Using recent advances in molecular biology, there is a growing potential for new classes of vaccines. The dissection of pathogens into their various components allows the development of specific subunit vaccines that are just as efficacious but are safer than whole pathogen vaccines. However, despite recent advances in vaccine research, the most common route of vaccination remains that of parenteral injection. The development of a broadly applicable oral delivery system remains a goal of the biotechnology industry for the efficient widespread administration of vaccines, but unfortunately this has proven impractical in most cases to date. The use of subunit vaccines for oral delivery has been generally resisted because of the obvious likelihood of protein degradation in the gut. Furthermore, even if the protein were to survive within an oral delivery system, there is no certainty that trafficking the protein to the gut would be sufficient to mount an immune response. Recently, transgenic plants have been investigated as an alternative means to produce and deliver vaccines. There are several reports demonstrating that antigens derived from various pathogens can be synthesized at high levels and in their authentic forms in plants (Arakawa *et al.*, 1997, Gomez *et al.*, 1998, Mason, *et al.*, 1992). When administered orally by feeding, such antigens can induce an immune response (Streatfield *et al.*, 2001, Haq *et al.*, 1995, Mason *et al.*, 1996) and, in some cases, result in protection against a subsequent challenge with the pathogen (ProdiGene, unpublished data, Arakawa *et al.*, 1997, Mason 1998). Certain antigens expressed in plants have shown sufficient promise to warrant human clinical trials (Tacket *et al.*, 1998, 2000). This has led to optimism that the inherent advantages of plants can be used to dramatically change the way in which vaccines can be delivered, and indeed that plants can become the delivery vehicle of choice for future vaccines. Combining the normal use of plants as human foods and as animal feed, with the production of vaccine subunit components in plant tissues, should allow vaccines to be produced competitively with the cost of other approaches.

A number of different plant systems have recently been under investigation for use as edible oral delivery systems. Of these, a system based on the use of transgenic maize seed appears to be the most realistic for a number of reasons. Among these reasons are the ability to introduce a grain-based product directly into a feed or food system, the ability to utilize the already existing infrastructure for the production, harvesting, transportation, storage, and processing of the grain, the ability to deliver a product (both monovalent and multivalent) at a cost competitive with contemporary vaccines due to a low cost of goods, and a plant system amenable to transformation with highly developed and characterized genetics. The use of corn grain is being explored as a particularly convenient delivery system for edible vaccines using a commercial animal example, a vaccine against swine transmissible gastroenteritis virus (TGEV). Additionally, a vaccine directed against enterotoxigenic strains of *Escherichia coli* (ETEC) is being developed as a model system to further develop vaccines for animal health. A major disease agent of ETEC is the heat-labile toxin (LT). This toxin has a multi-subunit structure very similar to cholera toxin and consists of a pentamer of receptor binding (B) subunits and a single enzymatic (A) subunit (Sixma *et al*., 1991). Approximately 66% of ETEC strains harbor LT, and in about half of these strains LT is the only toxin present (Svennerholm and Holmgren, 1995). ProdiGene has expressed LT-B in corn and demonstrated its immunogenicity and efficacy when fed to mice as a model system for the development of edible vaccines for animal health (Streatfield *et al*, 2001).

Swine transmissible gastroenteritis (TGE) (Saif and Wesley, 1992) is recognized as one of the major causes of sickness and death in piglets particularly in areas with high concentrations of pigs or regions with poor sanitation. TGE is a highly contagious enteric disease that is characterized by vomiting, severe diarrhea and high mortality in piglets less than two weeks of age. The causal agent of TGE is a pleomorphic, enveloped single-stranded RNA virus belonging to the genus *Coronavirus* of the family Coronaviridae. The virion contains three structural proteins designated M, N and S. Protein M is an integral membrane protein, N is a phosphoprotein that encapsulates the viral RNA genome, and S (spike) is a large surface glycoprotein (Laude *et al*., 1990) Replication of virus in the villous epithelial cells of the small intestine results in the destruction or alteration of function of these cells. These changes lead to a reduction in the activity of the small intestine that disrupts digestion and cellular transport of nutrients and electrolytes. In small piglets this can lead to a severe and fatal deprivation of nutrients and dehydration. Following infection, pigs that have survived the infection are immune to subsequent infections presumably due to local immunity in the intestinal mucosa. Thus, since active immunity towards TGEV involves local immunity, presumably through the activation and secretion of intestinal SIgA, edible vaccines that target activation of the intestinal mucosa immune system are particularly attractive in the control of this disease. ProdiGene has generated transgenic maize plants that express the spike protein at high levels. Corn expressing the S protein of TGEV was fed to 13-day-old piglets for ten days and subsequently challenged with a virulent Purdue strain of TGEV. This group of piglets was significantly protected from the disease in contrast to the control group that was fed non-transgenic corn. Results from a second trial duplicated these results (Streatfield *et al.*, 2001) demonstrating that the delivery of antigens in an edible oral form is efficacious.

Thus the development of edible vaccines offers the potential to aid in the control of enteric diseases such as ETEC and TGE. Edible vaccines from plant material could be directly delivered in the feed and could be produced cheaply in large volumes thus avoiding many costs associated with the administration of conventional vaccines. Vaccines from plants are particularly suitable for stimulation of mucosal immunity, since edible plant products can be delivered orally to reach the gut mucosal tissue and elicit an immune response at mucosal surfaces. Recent advances in technology make it now possible to express vaccine antigens at high levels in plants.

SELECTED EXAMPLES

LT-B corn fed to mice induces an immune response that combats LT holotoxin

It was first investigated whether LT-B produced in corn would induce an immune response when fed to mice. Mice were fed ground transgenic corn seed, then serum and fecal samples were analyzed for immune responses. Notably, equivalent amounts of pure LT-B or transgenic corn-expressed LT-B induce similar *anti*-LT-B specific Ig responses in serum (Fig. 1A). The response is clearly evident at 13 days after the first feeding and remains elevated for the course of the study. Doses of 5 mg of LT-B expressed in corn are sufficient to give a strong Ig response in serum, demonstrating that corn is an effective oral delivery vehicle for LT-B. As a guide to mucosal immunogenicity, *anti*-LT-B specific IgA levels were measured in fecal material of mice that had been fed LT-B expressed in corn. Responses are evident after 7 days and clearly cycle with peak responses about 1 week after each dose (Fig. 1B). As with the serum Ig response, doses of 5 mg of LT-B expressed in corn are sufficient to induce a strong mucosal IgA response. Strikingly, LT-B expressed in corn induces a much greater *anti*-LT-B specific mucosal IgA response than pure LT-B.

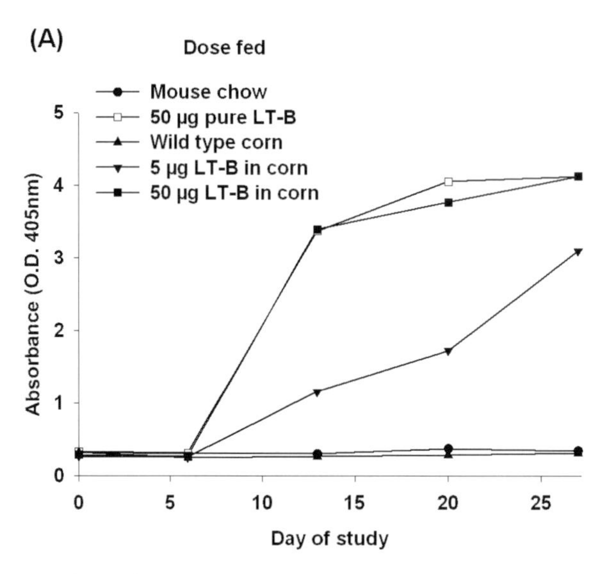

Figure 1A. *Protective immune responses of mice fed transgenic LT-B corn. Anti-LT-B specific Ig in serum. The mean response for the seven mice in each group is shown.*

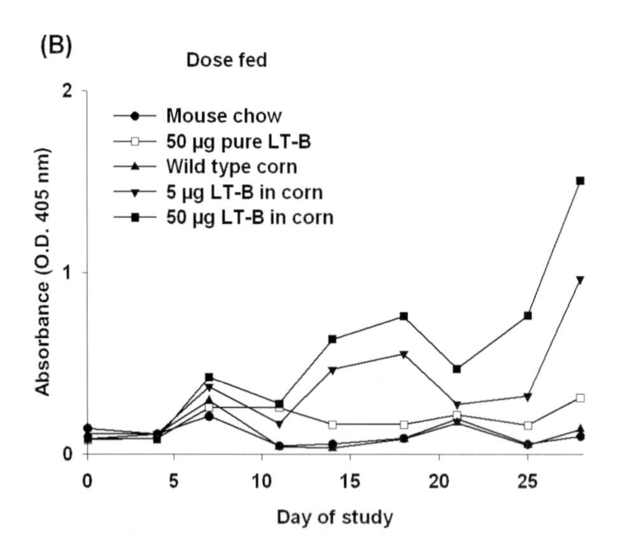

Figure 1B. *Anti-LT-B specific IgA in fecal material. The mean response for the seven mice in each group is shown; (C) The degree of gut swelling following challenge with LT holotoxin. Mean values for the weight ratios are shown with 95% confidence levels, and the sample size is given (n).*

In order to assess the efficacy of LT-B expressed in corn, LT-B corn was examined to determine if oral ingestion of LT-B corn could prevent gut swelling in mice exposed to the LT holotoxin. The upper intestines of a control group of mice swell when gavaged with LT, whereas those of mice fed LT-B expressed in corn do not swell when challenged with LT (Fig. 1C). Thus, LT-B expressed in corn appears to be protective against LT.

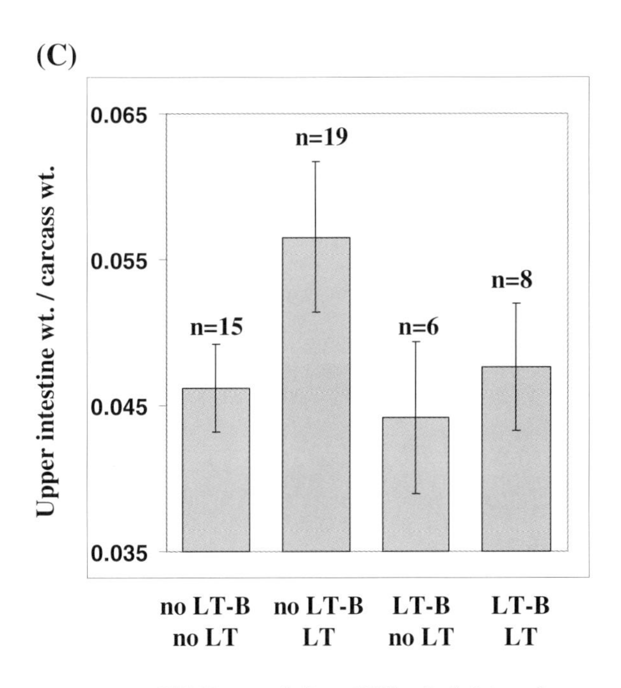

Figure 1C. *The degree of gut swelling following challenge with LT holotoxin. Mean values for the weight ratios are shown with 95% confidence levels, and the sample size is given (n).*

The S protein of TGEV expressed in corn is protective against viral infection.

Following the encouraging results with Lt-B, ProdiGene progressed to developing an edible vaccine against an economically important animal disease, TGE in swine. ProdiGene conducted a study to compare transgenic corn expressing the S protein of TGEV with a commercial modified live

TGEV vaccine. A negative control group fed wild type corn was also included. The percent morbidity incidence shows that all the piglets fed only wild type corn developed TGE clinical symptoms (Fig. 2A). Percent Morbidity Incidence was calculated as the number of animals with clinical signs > 2 divided by the total number of animals. Animals were monitored for clinical symptom development and scored on the appearance of various TGEV symptoms.

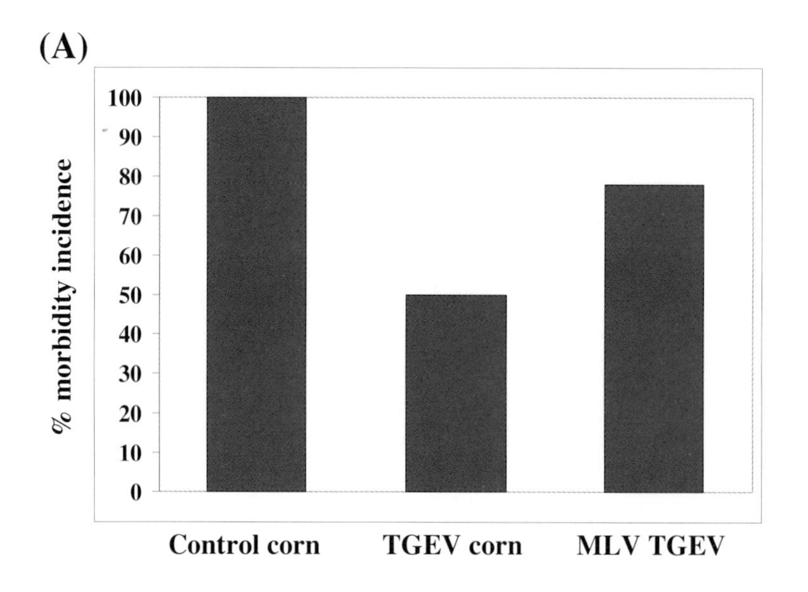

Figure 2A. *Percent morbidity incidence. Protection against TGEV of piglets fed transgenic corn expressing the S protein or modified live vaccine (MLV) TGEV.*

By comparison, only 50% of those animals that received the transgenic corn expressing the S protein exhibited symptoms. Interestingly, 78% of the piglets receiving the commercial modified live vaccine developed symptoms, indicating that the edible transgenic corn vaccine is more effective. However, when duration of symptoms is considered, (Fig. 2B) along with the clinical severity index, (Fig. 2C) it appeared that piglets that received the modified live vaccine recover as quickly as those that were fed transgenic corn expressing the S protein. Further studies with higher levels of orally delivered antigens should refine these results.

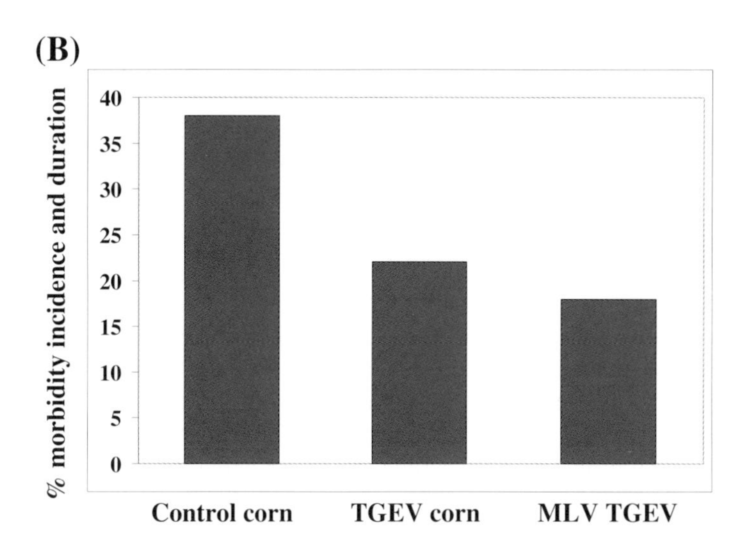

Figure 2B. *Percent morbidity incidence and duration*

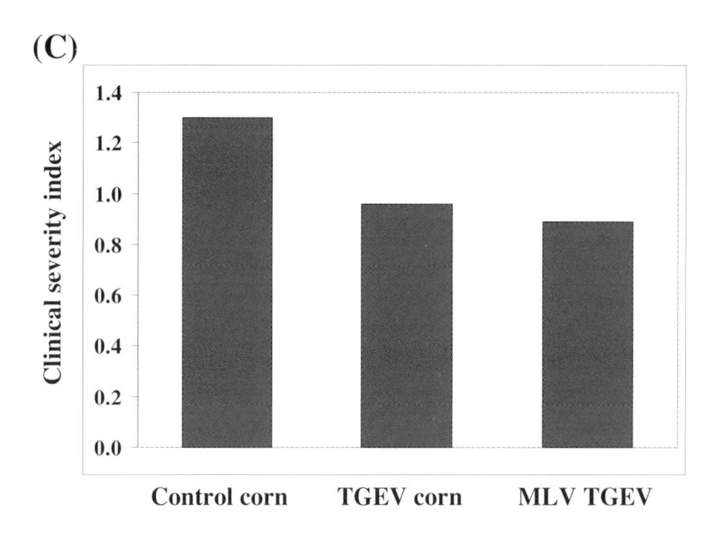

Figure 2C. *Clinical severity index. See text for definitions of clinical indices.*

Previously the full spike (S) protein was expressed in *Arabidopsis* (Gomez *et al.*, 1998). In this case, expression of the S protein was not detectable, yet plant extract injected intramuscularly into mice resulted in the production of detectable anti- S serum. More recently, the S protein has been expressed in tobacco (Tuboly *et al.*, 2000). The S protein was expressed at levels that could be detected by ELISA and a protein of the expected size was seen when analyzed by Western blotting. Leaf extracts from these plants were injected into 28-day old pigs. In comparison to those pigs that were injected with nontransgenic plant extract, measurable TGEV-specific antibodies were detected in the pigs. ProdiGene has extended these results by generating transgenic corn expressing the S protein that was fed to pigs in a virulent TGEV challenge study. This is the first demonstration of protection of an economically important animal from a naturally occurring disease by an oral vaccination using an edible system. Moreover this system uses conventional feed materials, e.g. corn, to deliver the antigen. One report (Modelska *et al.,* 1998) has shown in the laboratory the amelioration of rabies symptoms in mice fed multiple doses of a chimeric plant virus expressing the rabies glycoprotein following inoculation with an attenuated rabies strain. To our knowledge, until our report no animals in conventional food animal husbandry have been vaccinated with edible vaccines and shown to be protected from the disease. The level of protection seen in this study includes general health and vigor, a decrease in clinical symptoms, lack of virus shedding and other observations known to be criteria in disease protection. The mechanism of protection is unknown but may be an active immune response by the animal, competitive inhibition of viral receptor sites leading to non-establishment of a viral infection, or interference with parts of the viral replicative process.

THE FUTURE

Over the past decade, transgenic plants have been successfully used to express a variety of genes from bacterial and viral pathogens. Many of the resulting peptides induced an immunologic response in mice (Gomez *et al.*, 1998, Mason *et al.*, 1998, Wigdorovitz, *et al.*, 1999) and humans (Kapusta *et al.,* 1999) comparable to that of the original pathogen. Characterization studies of these engineered immunogens have proven the ability of plants to express, fold and modify proteins in a manner that is consistent with the authentic source.

Numerous genes have been cloned into a variety of transgenic plants including many enzymes that have demonstrated the same enzymatic activity as their authentic counterparts (Hood *et al.*, 1997, Hood and Woodard, this volume, Moldoveanu *et al.*, 1999, Trudel *et al.,* 1992). Animal and human immunization studies have demonstrated the effectiveness of many plant-derived recombinant antigens in stimulating the immune system. (See Table 1).

Table 1: *Examples of edible vaccines under development.*

Antigen Source	Plant System	Reference:
Bovine Pneumonic Pasteurellosis	White clover	**Lee RW** ***et al.*** 2001.
Cholera toxin	Tobacco	**Daniell H** ***et al.*** 2001.
B subunit of the Escherichia coli enterotoxin (recLT-B)	Potato	**Lauterslager TG** ***et al.*** 2001.
Hepatitis B surface antigen (HBsAg)	Potato	**Richter LJ** ***et al.*** 2000.
Respiratory syncytial virus (RSV)	Tomato	**Sandhu JS** ***et al.***2000.
Spike (S) protein of transmissible gastroenteritis virus (TGEV)	Tobacco	**Streatfield** ***et al.*** 2001 **Tuboly T** ***et al.*** 2000.
Hepatitis B virus surface antigen	Lupin (*Lupinus luteus* L.) and lettuce (*Lactuca sativa* L.) cv. Burpee Bibb	**Kapusta J** ***et al.*** 1999.
Foot and mouth disease virus (FMDV).	Alfalfa	**Wigdorovitz A** ***et al.*** 1999.
Rabies	Lettuce	**Modelska A** ***et al.***1998.
Hepatitis B surface antigen	Potato	**Tacket CO** ***et al.*** 1998.
B subunit of the Escherichia coli enterotoxin (recLT-B)	Potato	**Mason HS** ***et al.*** 1998.
Norwalk virus	Tobacco, potato	**Mason HS** ***et al.*** 1996.

Some of the first attempts to make edible vaccines included transgenic potatoes expressing the *E. coli* heat-labile enterotoxin (LT-B) (Haq *et al.*, 1995) and a Norwalk virus surface protein (Mason *et al.*, 1996). In both cases, mice fed the antigenic tubers produced serum and secretory antibodies specific to the authentic antigen. Subsequently, many plant-expressed antigens, including those referenced above, have been shown to elicit an immune response when administered through an oral route. Several of these antigens have shown sufficient promise to warrant human clinical trials (Mason *et al.*, 1998).

One of the most promising aspects of edible vaccines is the ability of orally administered immunogens to stimulate a mucosal immune response (Arakawa *et al.,* 1997) Mucosal surfaces, the linings of the respiratory, gastrointestinal, and urogenital tracts, play an important physical and chemical role in protecting the body from invading pathogens and harmful molecules. The mucosal immune system is distinct and independent of the systemic, or humoral, immune system, and is not effectively stimulated by parenteral administration of immunogens (Czerkinsky *et al* 1993). Rather, the mucosal immune system requires antigen presentation directly upon the mucosal surfaces. Since most invading pathogens first encounter one or more of the mucosal surfaces, stimulation of the mucosal immune system is often the best first defense against many transmissible diseases entering the body through oral, respiratory and urogenital routes (Holmgren *et al.*, 1994). Transgenic plants could produce large quantities of immunologically active recombinant antigen, very economically, for vaccine production. Multicomponent vaccines could easily be formulated from the seed of multiple transgenic plant lines to generate an increased chance for successful virus neutralization, in a stand-alone vaccination strategy, as a booster, or in combination with other vaccines and vaccination routes.

The use of transgenic grain for the delivery of animal healthcare based on preliminary results appears to have a legitimate place in the treatment and prevention of animal diseases. Such an approach will blur the line between traditional animal healthcare products and animal nutrition. Many parameters remain to be explored in the development of this technology. Some of these parameters include marketing and delivery of the product to the customer, the ability to fit the product into conventional husbandry techniques and the value and pricing of such products.

ACKNOWLEDGEMENTS

The amino acid sequences of the M, N and S proteins of an isolate of the Miller strain of TGEV were provided by Prem Paul, DVM (Iowa State University). The swine feeding trial was conducted at Ames, IA by Mark Welter (Oragen Technologies) and David Carter, DVM (Veterinary Resources, Inc.). John Clements (Tulane University) provided LT for the patent mouse assay. Leigh Anne Massey (ProdiGene) assisted in generating plant material for the mouse study and R. Craig Rainey (Texas A&M University) assisted in collecting mouse serum and fecal samples.

REFERENCES

Ahmad S, Lohman, B, Marthas M, Giavedoni L, el Amad, Z, Haigwood NL, Scandella CJ, Gardner MB, Luciw PA and Yilma T. 1994. Reduced virus load in rhesus macaques immunized with recombinant gp160 and challenged with simian immunodeficiency virus. AIDS Res. Hum. Retroviruses **10**:195-204.

Arakawa T, Chong DK, Merritt JL and Langridge WH. 1997. Expression of cholera toxin B subunit oligomers in transgenic potato plants. Transgenic Res. **6**:403-413.

Arakawa T, Yu J and Langridge WH. 1999. Food plant-delivered cholera toxin B subunit for vaccination and immunotolerization. Adv. Exp. Med. Biol. **464**:161-178.

Bergmeier LA, Mitchell EA, Hall G, Cranage MP, Cook N, Dennis M and Lehner T. 1998. Antibody-secreting cells specific for simian immunodeficiency virus antigens in lymphoid and mucosal tissues of immunized macaques. AIDS **12**:1139-1147.

Czerkinsky C, Svennerholm AM and Holmgren J. 1993. Induction and assessment of immunity at enteromucosal surfaces in humans: implications for vaccine development. Clin. Infect. Dis. 16 Suppl **2**:S106-S116.

Daniell H, Lee SB, Panchal T and Wiebe PO. 2001. Expression of the native cholera toxin b subunit gene and assembly as functional oligomers in transgenic tobaccochloroplasts. J Mol Biol ;**311**(5):1001-1009

Garvey PB, Hammond J, Dienelt MM, Hooper DC, Fu DF, Dietzschold B, Koprowski Hand Michaels FH. 1995. Expression of the rabies virus glycoprotein in transgenic tomatoes. Biotechnology (N. Y.) **13**:1484-1487.

Gomez N, Carrillo C, Salinas J, Parra F, Borca MV and Escribano JM. 1998. Expression of immunogenic glycoprotein S polypeptides from transmissible gastroenterities coronavirus in transgenic plants. Virology **249**:352-358.

Guidry JJ, Cardenas L, Cheng E and Clements JD. 1997. Role of receptor binding in toxicity, immunogenicity, and adjuvanticity of *Escherichia coli* heat-labile enterotoxin. Immunology 65:4943-4950.

Haq, TA, Mason, HS, Clements J and Arntzen CJ. 1995. Production of an orally immunogenic bacterial protein in transgenic plants: proof of concept of edible vaccines. Science 268:714-716.

Holmgren J, Czerkinsky C, Lycke N and Svennerholm AM. 1994. Strategies for the induction of immune responses at mucosal surfaces making use of cholera toxin B subunit as immunogen, carrier, and adjuvant. Am. J. Trop. Med. Hyg. **50**:42-54.

Hood, EE, Witcher DR, Maddock S, Meyer T, Baszczynski CBM, Flynn P, Register J, Marshal L, Bond D, Kulisek E, Kusnadi A, Evangelista R, Nikolov Z, Wooge C, Mehigh RJ, Hernan R, Kappel WK, Ritland, D, Li PC and Howard JA. 1997. Commercial production of avidin from transgenic maize: characterization of transformant, production, processing, extraction and purification. Molecular Breeding 3:291-306.

Hood EE and Jilka JM.1999 Plant-based production of xenogenic proteins. Curr Opin Biotechnol. 4:382-6

Ishida Y, Saito H, Ohta S, Hiei Y, Komari T and Kumashiro T. 1996. High efficiency transformation of maize (*Zea mays* L.) mediated by *Agrobacterium tumefaciens*. Nat. Biotechnol. 14:745-750.

Kapusta J, Modelska A, Figlerowicz M, Pniewski T, Letellier M, Lisowa O, Yusibov V, Koprowski H, Plucienniczak A and Legocki AB. 1999. A plant-derived edible vaccine against hepatitis B virus. FASEB J. **13**:1796-1799.

Kusnadi, A., R. Evangelista, E. E. Hood, J. A. Howard, and Z. Nikolov. 1998. Processing of transgenic corn seed and its effect on the recovery of recombinant ß-glucuronidase. Biotechnol. Bioeng. **60**:44-52.

Kusnadi AR. 1998. Production and purification of two recombinant proteins from transgenic corn. Biotechnol. Prog. 14:149-155.

Lauterslager TG, Florack DE, van der Wal TJ, Molthoff JW, Langeveld JP, Bosch D, Boersma WJ and Hilgers LA . 2001. Oral immunisation of naive and primed animals with transgenic potato tubers expressing LT-B. Vaccine **19**:2749-2755

Laude H, Rasschaert D, Delmas B, Godet M, Gelfi J and Charley B. 1990 Molecular biology of transmissible gastroenteritis virus. Vet Microbiol 23:147–54.

Lee RW, Strommer J, Hodgins D, Shewen PE and Niu Y,LoRY. 2001 Towards Development of an Edible Vaccine against Bovine Pneumonic Pasteurellosis Using Transgenic White Clover Expressing a Mannheimia haemolytica A1 Leukotoxin 50 Fusion Protein. Infect Immun 69(9):5786-5793.

Lu X, Kiyono H, Lu D, Kawabata S, Torten T, Srinivasan S, Dailey PJ, McGhee JR, Lehner T and CJ Miller. 1998. Targeted lymph-node immunization with whole inactivated simian immunodeficiency virus (SIV) or envelope and core subunit antigen vaccines does not reliably protect rhesus macaques from vaginal challenge with SIVmac251. AIDS 12:1-10.

Mason HS, Lam DM and Arntzen CJ. 1992 Expression of hepatitis B surface antigen in transgenic plants. Proc Natl Acad Sci USA;89:11 745–1 749.

Mason HS, Ball JM, Shi JJ, Jiang X, Estes MK and Arntzen CJ. 1996. Expression of Norwalk virus capsid protein in transgenic tobacco and potato and its oral immunogenicity in mice. Proc. Natl. Acad. Sci. U. S. A 93:5335-5340.

Mason HS, Haq TA, Clements JD and Arntzen CJ. 1998. Edible vaccine protects mice against Escherichia coli heat-labile enterotoxin (LT): potatoes expressing a synthetic LT-B gene. Vaccine **16**:1336-1343.

Modelska A, Dietzschold B, Sleysh N, Fu ZF, Steplewski K, Hooper DC, Koprowski H and Yusibov V. 1998 Immunization against rabies with plant-derived antigen. Proc Natl Acad Sci U S A Mar 3;95(5):2481-5.

Moldoveanu Z, Vzorov AN, Huang WQ, Mestecky J and Compans RW. 1999. Induction of immune responses to SIV antigens by mucosally administered vaccines [In Process Citation]. AIDS Res. Hum. Retroviruses **15**:1469-1476.

Pen J, Molendijk L, Quax WJ, Sijmons PC, van Ooyen AJ, van den Elzen PJ, Rietveld K and Hoekema A. 1992. Production of active Bacillus licheniformis alpha- amylase in tobacco and its application in starch liquefaction. Biotechnology (N. Y.) **10**:292-296.

Richter LJ, Thanavala Y, Arntzen CJ and Mason HS. 2000. Production of hepatitis B surface antigen in transgenic plants for oral immunization. Nat Biotechnol 18(11):1167-1171

Ruedl C and Wolf H. 1995. Features of oral immunization. Int. Arch. Allergy Immunol. 108:334-339.

Saif LJ and Wesley RD. 1992. Transmissible Gastroenteritis, Diseases of Swine 7th Edition, Iowa State University Press 29: 362-386.

Saif LJ, van Cott JL and Brim TA. 1994. Immunity to transmissible gastroenteritis virus and porcine respiratory coronavirus infections in swine. Vet. Immunol. Immunopathol. 43:89-97.

Sandhu JS, Krasnyanski SF, Domier LL, Korban SS, Osadjan MD and Buetow DE. 2000. Oral immunization of mice with transgenic tomato fruit expressing respiratory syncytial virus-F protein induces a systemic immune response. Transgenic Res 9(2):127-135.

Streatfield SJ, Jilka JM, Hood EE, Turner DD, Bailey MR, Mayor JM, Woodard SL, Beifuss KK, Horn ME, Delaney DE, Tizard IR and Howard JA. 2001. Plant-based vaccines: unique advantages. Vaccine(19):2742-2748.

Svennerholm A-M and Holmgren J. 1995 Oral B-subunit whole-cell vaccines against cholera and enterotoxigenic Escherichia coli diarrhea. In: Ala'Aldeen DAA, Hormaeche CE, editors. Molecular and Clinical Aspects of Bacterial Vaccine Development. Chichester, England: Wiley, 205–32.

Sixma TK, Pronk SE and Kalk KH *et al*. 1991. Crystal structure of a cholera toxin-related heat-labile enterotoxin from E. *col*i. Nature 351:371–377.

Tuboly T, Yu W, Bailey A, Degrandis S, Du S, Erickson L and Nagy E. 2000. Immunogenicity of porcine transmissible gastroenteritis virus spike protein expressed in plants. Vaccine (18):2023-2028.

Tacket CO, Mason HS, Losonsky G, Clements JD, Levine MM and Arntzen CJ. 1998. Immunogenicity in humans of a recombinant bacterial antigen delivered in a transgenic potato. Nat Med 4:607–9.

Tacket CO, Mason HS, Losonsky G, Estes MK, Levine MM and Arntzen CJ. 2000. Human immune responses to a novel Norwalk virus vaccine delivered in transgenic potatoes. J Infect Dis **182**:302–5.

Thanavala Y, Yang Y-F, Lyons P, Mason HS and Arntzen CJ. 1995. Immunogenicity of transgenic plant-derived hepatitis B surface antigen. Proc. Natl. Acad. Sci. U. S. A 92:3358-3361.

Trudel J, Potvin C and Asselin A. 1992. Expression of active hen egg white lysozyme in transgenic tobacco. Plant Sci. 87:55-67.

Tuboly T, Yu W, Bailey A, Degrandis S, Lerickson S Du and Nagy E. 2000. Immunogenicity of porcine transmissible gastroenteritis virus spike protein expressed in plants. Vaccine 18:2023-2028.

Wigdorovitz A, Carrillo C, Dus Santos MJ, Trono K, Peralta A, Gomez MC, Rios RD, Franzone PM, Sadir AM, Escribano JM and Borca MV. 1999. Induction of a protective antibody response to foot and mouth disease virus in mice following oral or parenteral immunization with alfalfa transgenic plants expressing the viral structural protein VP1. Virology **255**:347-353.

INDUSTRIAL PROTEINS PRODUCED FROM TRANSGENIC PLANTS

Elizabeth E. Hood and Susan L. Woodard

ProdiGene, Inc.
101 Gateway Boulevard
College Station, TX 77845 USA

INTRODUCTION

Industrial proteins include those used for industrial applications such as purification, diagnostics and enzymes for industrial processes. Enzymes as natural products are obtained from animal, plant and/or microbial tissues. When the industrial enzyme business began, enzymes were primarily obtained from plant and animal sources. However, today the bulk of commercial enzymes are derived from microbial sources either natural or recombinant. As a result, the extraction of commercial enzymes from plant and animal tissue has declined in recent years and continues to decline.

The advantages of using an enzyme as a catalyst in a particular process result from enzyme properties. Many enzymes operate at ambient temperatures and often in aqueous solution and usually at near neutral or physiological pH values. They have a very high specificity and function in low concentrations to produce the desired affect. Enzymes usually produce a comparatively rapid reaction and have a low level of toxicity. These characteristics allow industrial processes to occur in ambient conditions.

Unfortunately, enzymes are difficult and sometimes expensive to prepare, particularly in pure forms. They are generally unstable and require a degree of care and expertise in their use. In addition, extremes of pH and temperature can limit activity levels, even if these conditions do not destroy the enzymes. Enzymes may be inactivated by the presence of various ions, organic molecules or solvents, which can be present in organic reaction systems. The above adverse conditions are often present in large-scale industrial processes and significant changes to the process are often necessary when changing from chemical to enzymatic catalysts. Nevertheless, considering the large market opportunity for industrial enzymes and their environmentally friendly benefits, it is well worth the effort to generate an efficient production system, and make process changes to accommodate them.

E.E. Hood and J.A. Howard (eds.), Plants as Factories for Protein Production, 119–135.

Enzymes fall into several categories (Table 1), only a few of which are routinely utilized in industrial applications. Hydrolases are by far the most commonly used industrial enzymes. These enzymes, for example amylases and proteases, cleave molecules and in the process add water to the products. Other enzyme classes will be useful in industry when they become less expensive and thus available for testing in various processes. Oxidation/reduction (redox) enzymes are on the threshold of such a market entry.

Table 1. *Types of enzymes, their sources and their uses.*

Enzyme Class	% of Naturally Occurring	% of Dollar Value in Industrial Uses
Hydrolases	27	9
Transferases	27	Minor
Redox Enzymes	27	0.9
Lyases	11	0.2
Isomerases	4.5	1.7
Ligases	4.1	0

PROTEIN PRODUCTION SYSTEMS

Foreign protein production has been accomplished in several systems including bacteria, fungi, cultured animal cells, transgenic animals and of highest interest here, plants (Hood and Howard, 1999; Jilka *et al.*, 1999). Bacteria and fungi are relatively simple systems that are often employed. These microbial systems require a large capital outlay initially for fermentation equipment. Bacteria efficiently synthesize and secrete proteins and enzymes that are not glycosylated. The proteins may be easily purified if they remain soluble, however, insolubility is a problem encountered by many highly expressed proteins in bacterial systems. Fungi produce glycoproteins that can be secreted and consequently are relatively easy to purify. However, the post-translational processing may not be accurate in these systems—mis-folding of proteins in bacteria and atypical glycosylation of proteins in fungi may occur (Bowden *et al.*, 1991; Harashima, 1994; Archer, 1994). Immediate processing of material harvested from fermenters is also usually required.

Animal cell culture and transgenic animal production of foreign proteins are currently being explored in academic and industrial laboratories. Often, mammalian cell culture is used to produce monoclonal antibodies and viral vaccines. These systems are expensive and the scale up of transgenic animals is quite slow, making these systems most cost effective for high value proteins. The major advantage of these animal systems is that the sugars at the glycosylation sites more closely resemble those of the native protein in animals. Some pharmaceuticals, such as injectibles, may have this requirement.

THE PLANT PRODUCTION SYSTEM

Why use plants for production of industrial proteins and enzymes? Plants as a production system have many advantages over current competing technologies, particularly for industrial enzymes. In contrast to pharmaceuticals, the most important issue in the production of industrial enzymes is low cost. Plants are an excellent system for cost competition because of their protein expression potential and minimal production costs. A comparison of raw materials cost for different recombinant systems has been made (Table 2). These cost comparisons are based on assumptions for cost of capital equipment and maintenance of the cultures, animals or plants. (For a more complete treatment of this topic see the chapter in this volume by Nikolov *et al.*). Plant expression level must often be at least 0.1% of the dry weight of the production tissue to be cost effective. This level has been determined from cost models that factor in the cost of growing the crop, identity preservation of the seed or other plant part, shipping and whatever processing has to occur. This expression level has been attained for several proteins and enzymes (data not shown).

Table 2. *Raw material cost for proteins from recombinant systems.*

System	$ per gram recombinant protein
CHO* cells	300
Transgenic chickens/eggs	2.00
Transgenic goats/milk	2.00
Microbial fermentation	1.00
Plants	0.10

*Chinese Hamster Ovary

Several components can contribute to high expression levels. These include the obvious molecular components of good promoters and gene sequences that may need to be optimized for the plant of interest. However, targeting of the gene product to a particular tissue or subcellular space may have just as dramatic an affect on protein accumulation and stability as proper promoter choice. For example, enzymes expressed in the apoplast accumulate to much higher levels than many other locations within the cell (see trypsin and laccase data in later sections). This is in contrast to many studies of pharmaceutical protein expression in plants showing that the endoplasmic reticulum is the most lucrative expression site (Fiedler *et al.*, 1997). Another method of increasing expression of foreign proteins is through manipulation of germplasm. Takaiwa (2000) reported that crossing transgenic rice plants with mutant lines lacking the full complement of endosperm storage proteins improved the expression of the transgene under the control of an endosperm-preferred promoter. We also have shown affects on transgene expression through unusual germplasms (data not shown).

Plants generally offer a low cost of goods for manufacturing industrial enzymes for several reasons. Grain production is straightforward and involves normal agricultural practices. Production costs may contribute greatly to the overall cost of an industrial enzyme because of identity preservation, shipping and processing. Therefore, use of a commodity crop such as maize, soybeans or canola can help to control cost because of the sheer volume advantages of production. Also for some crops, production costs can be partially defrayed through by-product credits. An example of these credits is putting the solids from extraction of ground corn through an ethanol production facility (see Nikolov *et al.* in this volume). An additional cost advantage for crops is that scale-up time is abbreviated and inexpensive, mostly involving planting of increased acreage. Seed crops in particular contribute to cost control because proteins in seeds exhibit storage stability over many years (Z. Nikolov, personal communication) allowing production in favorable environments and consequential reliable supply of raw materials. Seeds can be processed to allow the direct application of the product to industrial processes. In addition, the products are safe to humans because of their lack of human pathogens.

However, we still have several challenges to address. First and foremost, for plant systems to compete with fermentation on similar products, the expression levels must be high enough to justify their use from an economic standpoint. Thus, the need for high expression drives technology development in this industry. Expression technology takes many forms. New promoters are being isolated from all plant systems being utilized today. These promoters take advantage of tissue-type specificity as well as preferred sinks in the plant or seed. Targeting experiments designed to take advantage of different subcellular compartments are being conducted. Gene optimization to address codon usage bias and remove RNA instability sequences is also often necessary to achieve high expression. Germplasm enhancement of gene expression is also a factor.

One of the biggest challenges we face in plant production of proteins, whether industrial proteins or vaccines, is containment in the field, and tracking the seed/plant through the production process. During the development of this industry, the players must take great care not to cross-contaminate other fields and seed storage areas or farm implements with any transgenic seed. This will ensure that products approved for industrial or medical use are not present in the food supply. A more complete treatment of this topic is contained in another chapter in this volume (Emlay).

EXAMPLES OF PLANT SYSTEMS

Plant systems that have been tested for enzyme production include tobacco, oilseeds, barley and maize (Table 3 and references therein). More recent work has been done in alfalfa (see chapter by Vezina *et al.*). Enzymes also have been expressed in *Arabidopsis* and tobacco tissue culture cells (Ziegler *et al.*, 2000), but these are experimental only, since they are not cost-effective production systems for industrial enzymes. In addition, some preliminary work has been done with potato (Table 3).

The earliest work was done on expression of phytase and alpha amylase in tobacco (Verwoerd *et al.*, 1995; Pen *et al.*, 1992). These studies certainly demonstrated the potential of plants as production systems (Table 3). Expression was recorded for leaf tissue, the most likely commercial source of protein from a plant such as tobacco. However, leaf tissue is not a good vehicle for stable storage of proteins. The proteins are not highly protected in leaf tissue, even when the tissue is dried. Therefore, in this system, proteins should be harvested immediately for the best recovery. Because cost is a major factor in the production of industrial enzymes, the expense of immediate extraction of fresh tobacco leaves makes it a less desirable production system than seeds. Recent studies (Ziegelhoffer *et al.*, 1999) with tobacco and with alfalfa, another leafy crop, have shown that cell wall degrading enzymes can be successfully expressed in these plants.

Oilseed crops are useful for the production of industrial enzymes (Table 3). Several plant species can be utilized to express proteins including canola, flax and safflower. These crops are relatively easy to transform (Moloney *et al.*, 1989) and produce as much as 1200 pounds of seed per acre. Expressing proteins on oil bodies assists in recovery (van Rooijen and Moloney, 1995; see chapter by Moloney *et al.*).

Maize, *Zea mays* L., is one of the most useful systems for production of proteins in plants, particularly industrial enzymes. The greatest advantage of maize is the scale-up potential and the low cost of production. To support scale up, the elaborate infrastructure that is in place to grow, harvest, store and process large volumes of corn seed can be exploited. Because industrial enzymes generally require a lower degree of purification for many applications, the seed crop as their carrier can add additional value when used in processes for food or as feedstock for industries that produce biological products. Biomass conversion and starch alteration are especially good targets. These advantages are different from those for plants used to produce therapeutics and vaccines. Non-animal sources of proteins and edibility of products are the drivers for these latter products-price is not the issue. However, for industrial enzymes, the bottom line is low cost and in most cases plants provide the low cost competitive system.

Table 3. *Expression details for xenogenic industrial proteins.*

Protein	Gene source	Plant	Expression Level	Promoter	Subcellular target	Comments	Reference
Phytase	Fungal (*Aspergillus*)	Tobacco	0.26% DW of leaves	CaMV 35S	Secreted	Not optimized	Verwoerd *et al.*, 1995
alpha amylase	Bacterial (*Bacillus*)	Tobacco	0.3% leaf TSP	CaMV 35S	Secreted	Glycosylated & still active	Pen *et al.*, 1992
Xylanase	Rumen fungus (*Neocallimastix*)	Canola	300-2000U per kg seed	Oleosin	Oil body surface	Fusion protein	Liu *et al.*, 1997
Xylanase	Rumen fungus (*Neocallimastix*)	Barley	0.004% DW of grain	GluB-1 (rice) Hor2-4 (barley)	Cytoplasmic	Second catalytic domain	Patel *et al.*, 2000
Xylanase	Fungal (*Clostridium*)	Tobacco	4% TSP of leaves	CaMV 35S	Secreted	Truncated xynZ Heat purified	Herbers *et al.*, 1995
Avidin	Chicken egg white	Corn	0.2% seed DW	Maize ubi-1	Secreted	Optimized gene	Hood *et al.*, 1997
Laccase	Fungal (*Trametes*)	Corn	0.005 to 0.025% DW of seed	Maize embryo-preferred	Secreted	Affects plant health	Hood *et al.*, PCT/US99/23256
Trypsin	Bovine pancreas	Corn	0.025% DW of seed	Maize embryo-preferred	Secreted	Successful when produced as a zymogen	S. Woodard *et al.*, this work; US Patent 6,087,558
1,4-ß-D-Endo-glucanase	Bacterial (*Acidothermus*)	Arabidopsis BY-2 cells	26% TSP of leaves	CaMV 35S	Secreted	Highly heat stable	Ziegler *et al.*, 2000
1,4-ß-D-Endo-glucanase	Bacterial (*Acidothermus*)	Potato	2.6% TSP of leaves	RbcS-3C Hybrid 35S enhancer + MAS*	Secreted or chloroplast targets	CaMV enhancer	Dai *et al.*, 2000
1,4-ß-D-Endo-glucanase	Bacterial (*Thermomonospora*)	Alfalfa Tobacco	14 μg/g DW (~0.01% TSP of leaves)** 0.1% TSP of leaves	Hybrid 35S enhancer + MAS*	Cytoplasmic	Bacterial leader removed	Ziegelhoffer *et al.*, 1999
Cellobio-hydrolase	Bacterial (*T. fusca*)	Alfalfa Tobacco	0.002% TSP of leaves 0.02% TSP of leaves	Hybrid 35S enhancer + MAS*	Cytoplasmic	Bacterial leader removed	Ziegelhoffer *et al.*, 1999
1,3-1,4-ß-D-glucanase	Hybrid of barley and *Bacillus* enzymes	Barley	Not stated	High pI barley alpha amylase	Secreted	Heat stable region of *Bacillus* enzyme added	Jensen *et al.*, 1996

**Agrobacterium*-derived mannopine synthase
**Assuming 10% of leaf weight is soluble protein.

EXAMPLES OF PROJECTS

In direct contrast to pharmaceutical proteins, functionality in an industrial process is the most important issue for industrial enzymes. Thus, whatever is necessary for the genes to express at high levels in transgenic plants--such as added KDEL and other targeting sequences--is more important than protein identity with its native source, as long as the enzymes are active. This also extends to glycosylation patterns. The following examples shed light on how high expression of industrial enzymes has been achieved.

The first commercial proteins

The first xenogenic proteins produced and sold from transgenic plants 1998 include avidin and ß-glucuronidase (GUS) (Hood *et al.*, 1999; Witcher *et al.*,). Avidin, the first plant-produced transgenic protein product to be marketed, was first sold in 1997. One of the prominent concerns for transgenically-produced proteins is stability of expression over time. After 10 generations in the field, expression levels for avidin have not only been maintained, but are increasing through selection and improvements in germplasm. Up to 2 g avidin can be extracted per kg of seed from ProdiGene's current production line. In the plant, high levels of avidin expressed from a constitutive promoter cause male sterility (Hood *et al.*, 1997), an advantage from a regulatory standpoint. Applications of these proteins include using the GUS protein as a research reagent and avidin as a diagnostic reagent and protein purification tool for biotinylated proteins.

Non-food industrial enzymes

Currently, the highest profile plant-based industrial enzyme projects underway involve enzymes for applications in feed, cleaning agents, processing reagents, the wood products industry and biomass conversion. Primarily these include xylanases and cellulases for biomass conversion and the wood products industry, as well as laccase and trypsin (see Table 3 and references therein) for some of the other applications. It is important to explore many plant systems, enzymes, promoters and targeting sequences to understand the factors that affect expression levels and hence the economics of industrial enzyme production in plants. To date, this experimentation has been limited.

The xylanases showed reasonable expression in plants (Table 3), although direct comparison of their expression levels is not possible among these plant systems because the amounts of enzyme reported in each case are not in equivalent units. The 35S promoter was used in tobacco for leaf expression, whereas seed expression was driven by endosperm promoters in barley and by an oleosin promoter in canola (Table 3). The enzyme was also expressed in different subcellular compartments: cell wall in tobacco, oil bodies in canola and cytoplasm in barley. Each of the differing components: promoters, subcellular compartments, tissue types and plant hosts, have been shown to have an impact on transgene expression. Therefore, without conversion factors not only for units to grams of enzyme but also among plant and tissue types, it is difficult to conclude which expression system worked the best.

For the cellulases, 1,4- ß-D-endoglucanase and cellobiohydrolase, comparing expression data was straightforward (Table 3). Assuming that 10% of the dry weight of alfalfa leaves is soluble protein, the expression in alfalfa leaves is approximately 0.01% total soluble protein (TSP), compared to tobacco at 0.1% TSP (Table 3; Ziegelhoffer *et al.*, 1999). In both cases the enzyme (E1, truncated form of a cellulase) is expressed with the CaMV 35S enhancer added to the mannopine synthase promoter from *Agrobacterium* and localized in the cytoplasm. These same authors demonstrated that cellobiohydrolase, an exoglucanase, was expressed approximately at 5-fold lower levels in leaves of both plant types. In contrast, the same promoter produced potato plants with the E1 cellulase expressed at 1.0% TSP in leaves. The major difference here is that the protein was secreted into the cell wall space (Table 3). Expression in potato leaves was somewhat higher, though not significantly so, when the gene was expressed from the promoter for the small subunit of ribulose bisphosphate carboxylase (rubisco) gene and targeted to the chloroplast (Dai *et al.*, 2000). The best expression, however, was achieved in *Arabidopsis* leaves (26% TSP) using the CaMV 35S promoter (Table 3; Ziegler *et al.*, 2000). In this case also the protein was secreted into the cell wall space. *Arabidopsis* is in the family of native species affected by CaMV and perhaps this has some impact on promoter function—i.e., induction of high transcription rates. Moreover, the E1 protein expressed in these plants was a highly truncated version of a gene from a different organism than those discussed above. The gene fragment used emphasizes the thermo-stable catalytic subunit, which may have a more dramatic affect on expression than expressing the holoprotein from a stronger promoter in other plant systems. Additionally, perhaps the cell wall is a more stable environment than the cytoplasm or the chloroplast for the accumulation of the cellulase proteins. We have found the cell wall space to be the best location for accumulation of many types of proteins (Hood *et al.*, 1997; Zhong *et al.*, 1999; this work; E. Hood, unpublished).

Laccase

Laccase is a blue copper oxidase that acts through the generation of free radicals by extracting electrons from target substrates and donating them to molecular oxygen. The free radicals generated then react with other molecules to form or degrade polymers. Because these enzymes have a major impact on cellular metabolism, care must be taken to express these enzymes in a developmentally and spatially protective manner. The most effective means of achieving high level expression consists of using a seed-preferred promoter and targeting the enzyme to the cell wall space (PCT/US99/23256).

We expressed the laccase 1 isozyme from *Trametes versicolor* in transgenic maize using constitutive and seed-specific promoters (Table 4). Several transgenic events were recovered from the constitutive and embryo-preferred promoter vectors and one for the endosperm-preferred promoter vector. Expression levels in T1 seed were quite dramatically different for events from these vectors, with the embryo-preferred promoter and cell wall targeting, out-performing all other vector combinations. When this same promoter was used to target laccase to other subcellular compartments, far lower expression was seen, suggesting that this enzyme is most favorably stored in the apoplastic compartment. T1 expression levels in single seed (seed from the first transgenic generation; Table 4) were determined by enzyme assay after incubating the extracts with copper salts. The line contained ~2.5-5 ng laccase per mg dry weight of seed, an amount that correlated well with the amount of protein detected on a Western blot (Fig. 1). Field studies have yielded bulk seed populations which express extractable laccase at 50 ng/mg. These bulk populations contain seed that is negative for laccase as well as seed that is positive with a single gene dose (i.e. is hemizygous for the transgenic locus), indicating that the positive seed have even higher levels of extractable laccase. The expression level of laccase necessary for cost-effective commercial production is approximately 3-10 fold greater than levels for the lines described above for an extracted product, and can be achieved through breeding and selection.

Table 4. *Laccase expression vectors and events recovered*

Vector Name	Promoter	Subcellular Target	Gene	# Events	T1 expr. levels*
LCD	Const.	CW	Laccase	15	~1-2 ng/mg
LCG	Embryo	CW	Laccase	18	~2.5-5 ng/mg
LCH	Endosperm	ER	Laccase	1	~0.1-0.2 ng/mg

*Expression level was determined using an enzyme assay. The level reported is the highest T1 seeds identified from each construct.

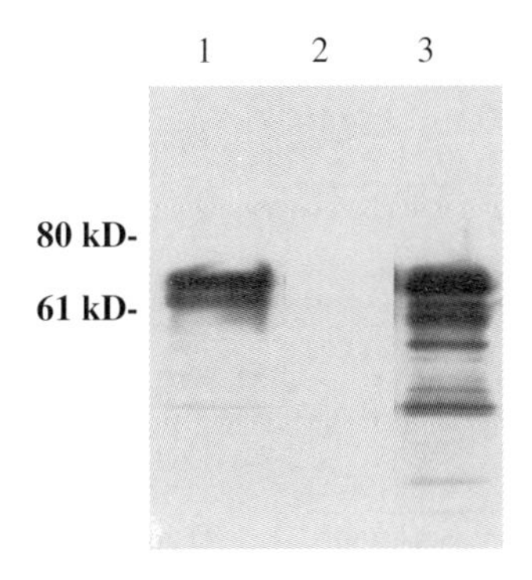

Figure 1. *Western blot of corn seed extract containing laccase. Lane 1, 10 ng of recombinant Trametes laccase 1 from fungal broth; Lane 2, 1 μg of negative control corn seed extract; Lane 3, 1 μg of LCG corn seed extract.*

Trypsin

Trypsin is a pancreatic enzyme involved in food digestion. We have expressed the bovine pancreatic trypsin in maize lines for large scale production for industrial applications. Proteases in general are difficult to produce in xenogenic systems because of their detrimental affect on native protein content. ProdiGene was recently issued a broad-based patent (USP # 6,087,558) directed toward the production of proteases in transgenic plants, claiming expression of any protease in any transgenic plant, where the protease is expressed in the zymogen form. This is significant, because recovery of transgenic plants expressing high levels of active proteases is difficult to impossible without expressing zymogens in seed.

Events from constitutively expressed mature trypsin were recovered with a much lower frequency than events containing the zymogen expressed under a constitutive promoter (data not shown). In addition, expression levels for events containing active trypsin were below or near the lower limit of detection in our assay while expression levels for trypsinogen were much higher (Table 5). Events for trypsinogen expressed with an embryo-preferred promoter were recovered with the same frequency as those for trypsinogen expressed constitutively, though enzyme levels for the seed-targeted trypsinogen were much higher (Table 5). These results support our claim that seed-preferred expression of the zymogen gives the highest expression of a protease in transgenic plants. As was the case for laccase, the best subcellular location for accumulation of this enzyme is the apoplast or cell wall (Table 5).

Table 5. *Trypsin constructs and events recovered.*

Vector Name	Promoter	Subcellular Target	Gene	No. of events	T1 expr. Levels*
TRC	Constitutive	Cell wall	Trypsinogen	15	0.057% TSP
TRD	Constitutive	Cell wall	Trypsin	5	0.010% TSP
TRE	Endosperm	Amyloplast	Trypsinogen	10	0.75% TSP
TRF	Embryo	Cell wall	Trypsinogen	16	3.3% TSP

*Expression level was determined using an enzyme assay. The level reported is the highest T1 seed identified from each construct.

Since trypsinogen itself has a low level of enzyme activity and is autocatalytic (Kunitz and Northrop, 1937), it was not surprising to find trypsinogen as well as active trypsin in the seed. In order to determine the best way to estimate recombinant trypsin concentrations, we compared various assays used to determine expression levels from seeds of our best lines (Table 6). The ELISA underestimates the level of protein made when a high level of active enzyme is produced. This phenomenon is probably due to proteolytic digestion of the antibodies used in the assay. In spite of the fact that pro- and active forms of the enzyme are present in the extracts, the expression levels determined by enzyme assay compare more closely with the results obtained by Western blot analysis. Neither of these assay methods compares favorably with the ELISA.

Table 6. *Comparison of trypsin recovered as a percent of total soluble protein obtained by different methods of analysis.*

Sample identifier	ELISA	Enzyme assay	Western blot*
TRF11060-5	0.10	0.92	1.25
TRF13070-1	0.073	1.5	2.0
TRF15070-5	0.090	3.3	>2.5

*Western blot results estimated to nearest 0.25%.

We show a representative Western blot for trypsinogen expression in different seeds (Fig. 2). The five seeds shown were identified as positive in a trypsin activity assay, then run on a polyacrylamide gel which was used to generate a Western blot. The blot was developed using an antibody raised against bovine trypsinogen. This result shows that several protein bands that cross-react with the anti-trypsinogen antibodies are present in the corn seed extract. More than one of these shows activity on a zymogram (data not shown). Experiments are in progress to determine the source of these molecular weight variants. In addition, we are continuing to examine promoter and targeting sequence combinations that increase expression further. High-expressing TRF lines are currently being crossed into elite inbreds to increase yield and expression levels.

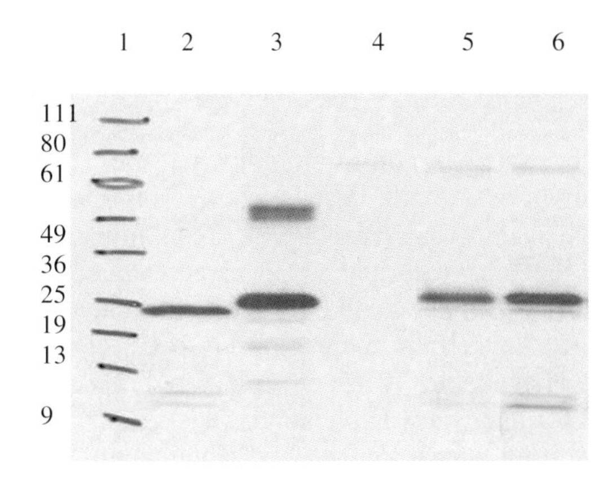

Figure 2. *Western blot of transgenic lines expressing trypsinogen/trypsin. Lane 1 shows the position of molecular weight markers; Lane 2, 5 ng of pure trypsin; Lane 3, 5 ng of pure trypsinogen; Lane 4, 2 μg of negative control corn seed extract; Lane 5, 2 μg of a representative TRE seed extract; Lane 6, 2 μg of a representative TRF seed extract.*

Trypsin has applications in cell culture and protein processing, often for production of proteins for pharmaceutical use. Currently the enzyme is being produced from bovine refuse from slaughterhouses. With the existing problems arising from hoof and mouth disease and mad cow disease (BSE), non-animal sources of the enzyme are in high demand. Trypsin produced from transgenic maize can meet this demand.

POSSIBILITIES FOR NEW PRODUCTS

Foods

In the food area, enzymes are used in the conversion of raw materials to form intermediate products more useful in food processing and in food formulations. The treatment of food products with enzymes makes them more palatable, more stable or enables the development of some other desirable property. Enzyme production for the food industry is primarily through microbial fermentation. However, major value from edible plant-based protein or enzyme production is readily apparent. For example, ProdiGene has expressed the sweet protein, brazzein, in transgenic maize. For baking applications, the direct addition of corn meal is possible, saving major dollars in processing costs.

Non-food industrial uses

Currently the principal non-food industrial uses for enzymes are starch hydrolysis (amylases), textile desizing (amylases), leather production (proteases), detergent additives (proteases), and animal feeds. Starch hydrolysis and detergent enzymes and, to a lesser extent, textile enzymes, represent the only substantial markets to date. However, one of the reasons for limited use of enzymes in certain applications is because no cost-effective source of the enzyme is available. Transgenic plant systems can meet both the scale and cost targets for many new applications.

New and exciting areas for enzyme applications are in biomass conversion and the wood products industries. For biomass conversion applications, enzymes that degrade cell walls will be needed. The resulting products from those enzymes are monomeric components of walls, primarily 5 and 6 carbon sugars. These will be substrates for a variety of applications including fermentation into ethanol. Oxidation/reduction enzymes such as laccase (described above) and peroxidases (de Jong *et al.*, 1992; Jansen *et al.*, 1992; Caramelo *et al.*, 1999) will find many new markets when high level expression of these enzymes is achieved in any system. These enzymes can potentially replace many applications currently using chemicals that are damaging to the environment. Because of the scale-up potential of the plant system and the low cost of goods, plants will likely be the system of choice for industrial scale enzyme production for these industries. A comparison of the speed and potential of scale-up for several transgenic systems is instructive (Table 7). It is clear that for large-scale production, transgenic animals and animal cell culture (CHO cells) are not practical. It is also clear that plants have the shortest scale-up period, and production timelines after line development are quite abbreviated. Plant molecular farming is an exciting new area and the potential is enormous.

Table 7. *Comparison of scale-up potential for several recombinant protein production systems, starting with material containing 1 gram of active ingredient.*

	Time to achieve indicated quantities		
System	**200 g**	**40 kg**	**8000 kg**
Plants	4 months	8 months	12 months
Transgenic animals	6 months	3 years	Not practical
Fermentation	4 months	3 years	3 years
CHO cells	18 months	3 years	Not practical

REFERENCES

Archer DB. 1994. Enzyme production by recombinant *Aspergillus*. Bioprocess Technol. 19:373-393.

Bowden GA, Paredes AM and Georgious G. 1991. Structure and morphology of protein inclusion bodies in Escherichia coli. Biotechnology 9:725-730.

Caramelo L, Martinez MJ and Martinez AT. 1999. A search for lignolytic peroxidases in the fungus *Pleurotus eryngii* involving alpha-keto-gamma thiomethylbutyric acid and lignin model dimers. Appl. Env. Micro. 65:916-922.

Dai Z, Hooker BS, Anderson DB and Thomas SR. 2000. Improved plant-based production of E1 endoglucanase using potato: expression optimization and tissue targeting. *Molecular Breeding* 6: 277-285.

DeJong E, de Vries FP, Field JA, van der Zwan RP and de Bont JAM. 1992. Isolation and screening of basidiomycetes with high peroxidative activity. Mycol. Res. 96:1098-1104.

Fiedler U, Phillips J, Artsaenko O and Conrad U. 1997. Optimization of scFv antibody production in transgenic plants. Immunotechology 3:205-216.

Harashima S. 1994. Heterologous protein production by yeast host-vector systems. Bioprocess Techol. 19:137-158.

Herbers K, Wilke I and Sonnewald U. 1995. A thermostable xylanase from Clostridium thermocellum expressed at high levels in the apoplast of transgenic tobacco has no detrimental effects and is easily purified. *Bio/technology, Bio/technology.* 13: 63-66.

Hood E and Howard J. 1999. Protein products from transgenic plants. *Agro-Food-Industry Hi-Tech,* 3, 10: 35-36

Hood EE, Kusnadi A, Nikolov Z and Howard, JA. 1999. Molecular farming of industrial proteins from transgenic maize. *Chemicals via Higher Plant Bioengineering,* pp. 127-147.

Hood EE and Jilka JM. 1999. Plant-based production of xenogenic proteins. *Current Opinion in Biotechnology* 10: 382-386.

Hood EE, Witcher DR, Maddock S, Meyer T, Baszczynski C, Bailey M, Flynn P, Register J, Marshall L, Bond D, Kulisek E, Kusnadi A, Evangelista R, Nikolov Z, Wooge C, Mehigh RJ, Hernan R, Kappel WK, Ritland D, Li CP and Howard JA. 1997. Commercial production of avidin from transgenic maize: Characterization of transformant, production, processing, extraction and purification. *Molecular Breeding* 3: 291-306.

Janse BJH, Gaskell J, Akhtar M and Cullen D. 1998. Expression of *Phanerochaete chrysosporium* genes encoding lignin peroxidases, manganese peroxidases, and glyoxal oxidase in wood. *Appl. Env. Microbiol.* 64:3536-3538.

Jensen LG, Olsen O, Kops O, Wolf N, Thomsen KK and von Wettstein D. 1996. Transgenic barley expressing a protein-engineered, thermostable (1,3-1,4)-β-glucanase during germination. *Proc. Natl. Acad. Sci. USA* 93: 3487-3491.

Jilka J, Hood EE, Dose R and Howard J. 1999. The benefits of proteins produced in transgenic plants. *AgBiotechNet.* 1:1-4.

Kunitz M and Northrop JH. 1936. Isolation from beef pancreas of crystalline trypsinogen, trypsin, a trypsin inhibitor, and an inhibitor-trypsin compound. *J. Gen. Physiol.,* 19: 991-1007.

Leite A, Kemper EL, da Silva MJ, Luchessi AD, Siloto RMP, Bonaccorsi ED, El-Dorry HF and Arruda P. 2000. Expression of correctly processed human growth hormone in seeds of transgenic tobacco plants. *Molecular Breeding* 6: 47-53.

Liu J-H, Selinger LB, Cheng K-J, Beauchemin KA and Moloney MM. 1997. Plant seed oil-bodies as an immobilization matrix for a recombinant xylanase from the rumen fungus *Neocallimastix patriciarum*. *Molecular Breeding* 3: 463-470.

Moloney MM, Walker JM and Sharma KK. 1989. High efficiency transformation of *Brassica napus* using *Agrobacterium* vectors. *Plant Cell Reports* 8: 238-242.

Patel M, Johnson JS, Brettell RIS, Jacobsen J and Xue G-P. 2000. Transgenic barley expressing a fungal xylanase gene in the endosperm of the developing grains. *Molecular Breeding* 6: 113-123.

Pen J, Molendijk L, Quax WJ, Sijmons PC, van Ooyen AJJ, van den Elzen PJM, Rietveld K and Hoekema A. 1992. Production of active *bacillus licheniformis* alpha-amylase in tobacco and its application in starch liquefaction. *Bio/Technol*, 10:292-296.

Takaiwa F, Tada Y, Wu C-Y, Washida H and Utsumi S. 2000. High level accumulation of soybean glycinin in rice endosperm. *Plant Mol. Biol. Rptr. Supplement* 18:2: S28-28.

Van Rooijen GJH and Moloney M. 1995. Plant seed oil-bodies as carriers for foreign proteins. *BioTechnology*, 13: 72-77.

Verwoerd TC, van Paridon PA, van Ooyen AJJ, van Lent JWM, Hoekema A and Pen J. 1995. Stable accumulation of *Aspergillus niger* phytase in transgenic tobacco leaves. *Plant Physiol.* 109: 1199-1205.

Witcher DR, Hood EE, Peterson D, Bailey M, Bond D, Kusnadi A, Evangelista R, Nikolov Z, Wooge C, Mehigh R, Kappel W, Register JC and Howard J. 1998. Commercial production of β-glucuronidase (GUS): A model system for the production of proteins in plants. *Molecular Breeding* 4: 301-312.

Zhong GY, Peterson D, Delaney DE, Bailey M, Witcher DR, Register III JC, Bond D, Li C-P, Marshall L, Kulisek E, Ritland D, Meyer T, Hood EE and Howard JA. Commercial production of aprotinin in transgenic maize seeds. *Molecular Breeding_* **5**: 345-356, 1999.

Ziegelhoffer T, Will J, and Austin-Phillips S. 1999. Expression of bacterial cellulase genes in transgenic alfalfa (*Medicago sativa* L.), potato (*Solanum tuberosum* L.) and tobacco (*Nicotiana tabacum* L.). *Molecular Breeding* 5: 309-318.

Ziegler MT, Thomas SR, and Danna KJ. 2000. Accumulation of a thermostable endo-1,4-ß-D-glucanase in the apoplast of *Arabidopsis thaliana* leaves. *Molecular Breeding* 6: 37-46.

Part III
Production Issues

CHOICE OF CROP SPECIES AND DEVELOPMENT OF TRANSGENIC PRODUCT LINES

Donna E. Delaney

ProdiGene, Inc.
101 Gateway Boulevard
College Station, TX 77845

INTRODUCTION

The choice of a crop species for use as a manufacturing vehicle for heterologous protein production is a serious one because it will affect every aspect of product development from the construction of transformation vectors to the downstream processing steps necessary for purification. The use of some crops (e.g. soybeans and canola is further limited by patent restrictions on commercial transformation. Transformation of any crop species will generate an enormous amount of variation in recombinant protein expression and agronomic characteristics that must be managed by careful selection and breeding. For some crops this may be a fairly simple matter, whereas for others it may require years to obtain a finished product line. The breeding of transgenic crops has its own unique challenges and should be approached differently than standard plant breeding because the objectives may be very different depending on the product and target market.

CHOICE OF CROP SPECIES

When choosing a crop for the manufacture of heterologous proteins there are many factors to consider. The optimal system would offer flexibility, ease of manipulation, high protein accumulation, rapid scale-up and low production and handling costs. The type of product and size of the potential market may also influence the choice. For high value, low volume products, such as some pharmaceuticals, many different crops could be functional and economical production systems. However, for products where volume and low cost production are critical for market entry, such as industrial enzymes, only the major crop species with well established infrastructures for production, handling and transport will suffice. In the end,

E.E. Hood and J.A. Howard (eds.), Plants as Factories for Protein Production, 139–158.

one has to decide whether to choose a system that would allow the production of many types of products, or whether to specialize.

Amenability to transformation

One of the major factors that influences the speed of product development is the ability to transform and regenerate elite cultivars or inbreds. Nearly every crop species can be stably transformed, but for many the system only works efficiently with one or a few select lines, which are often not the most current elite germplasm, and some are not suitable for field production. Since plants produced for heterologous protein production are generally not intended for the seed industry, for some high value products this may not be a significant problem. Any plant that produces the desired protein in a useful form will suffice. However, for products which are high volume and for which production costs must be kept to a minimum, a high yielding product line will be critical. For the Solanaceous crops (tobacco, tomato and potato), for example, elite cultivars are readily transformable, while for many other species only a limited number of lines can be routinely used. In the latter case, the gene of interest must then be transferred to elite germplasm using standard plant breeding techniques and may require years to develop a finished product line. Recombinant protein production can continue throughout this development phase if one is willing to accept the higher production costs that may be incurred from use of an inferior plant genotype.

Ease of genetic manipulation and agronomic improvement

Regardless of whether the transformation was performed on elite germplasm, all newly transformed (T0 and T1) lines will require some degree of genetic improvement to produce a useful production line. Vegetatively propagated crops, such as potato, may only require some selection among clones for agronomic type and high expression of the desired protein. For sexually propagated crops the development time will vary widely depending on the transformation system used, pollination method (self pollinated vs. cross pollinated), and type of production line (inbred, hybrid or synthetic). A generation or two of inbreeding and selection may be sufficient for some crops, while for others it may be necessary to perform several generations of backcrossing, selection, inbreeding and hybrid production. Methods of artificial pollination have been established for all crops, but this process is much easier in some crops than in others. The ease of this process, and the number of seeds produced per pollination, will affect development costs and scale-up efficiency. A longer development phase may not necessarily be a disadvantage since it increases the opportunities for selection of stable, high expressing lines, which will increase the success and longevity of the product line.

Polyploid crops such as potato, alfalfa and wheat pose special problems for transgene expression because of the complex nature of their genomes. Autopolyploid crops are often highly heterozygous and extremely averse to inbreeding. The high degree of heterozygosity has been suggested to be maintained not only by outcrossing, but also by a concerted effort by the plant genome to silence and/or mutate multiple copies of a gene, in effect an attempt to "diploidize" the genome (Mittelsten Scheid *et al.* 1996; Comai 2000). This makes it very difficult to produce non-segregating lines and increases the frequency of gene silencing events. If the transformation is performed at the diploid level, and then the ploidy is increased (by chromosome doubling or crossing to a tetraploid), this can also lead to gene silencing and/or suppression of transgene expression (Mittelsten Scheid *et al.* 1996). Clonal or vegetative propagation of the transgenic line may alleviate some of these problems, however further problems may occur once the crop is moved into field production. In potato, it was found that field grown tubers were more "polyploidized" than either microtubers from culture or from greenhouse grown plants (Hovenkamp-Hermelink *et al.* 1988); (Wolters & Visser 2000). This can lead to increased copy number of the transgene, which can trigger gene silencing. Selfing or inbreeding to produce homozygous lines will also increase the number of copies or "alleles" of the transgene, and this again will increase the frequency of gene silencing. It will also lead to inbreeding depression and poor field performance. Inbreeding is not a problem for diploidized allopolyploids such as wheat and oats, but the complex nature of these genomes and the difficulty in obtaining transformed plants can lead to some of the same problems. These crops are generally transformed by particle bombardment, which frequently produces transformants with multiple copies, multiple loci and rearrangements of the transgene (Pawlowski *et al.* 1998; Srivastava *et al.* 1999). All of these situations can lead to gene silencing and unstable expression.

Genetic manipulation of diploid crops is more straightforward and has fewer complications. Single or low copy transformants are more easily obtained and are stably inherited. However, there are advantages to some crops over others in terms of scalability, storage longevity and production costs. These factors are discussed below.

Scale-up potential

Compared to other methods of recombinant protein production, such as bacterial or mammalian cell culture or transgenic animals, the ease with which one can scale-up production of transgenic plants is a significant advantage. It is as simple as planting additional acres and requires minimal added cost to accomplish. While all plant systems can be scaled-up faster and more economically than fermentation facilities, there are some differences in speed of early scale-up (pre-log phase) among the different crop species, especially if any genetic improvements must be accomplished first. Tobacco probably has the highest multiplication capacity of any of the crops commonly used for protein production (Table 1). Hand pollinations typically yield up to 200 seeds per capsule and if successive harvests are made, up to 1 million seeds per plant have been reported (Cramer *et al.* 1999). Production of corn and sunflower can also be ramped up quickly because of the large numbers of seeds produced per pollination. Crops like soybeans however, where pollination procedures are tedious and yield very few seeds, will take several generations to ramp up to full production quantities.

Table 1. *Scale -up efficiency of different crop species*

Crop	No. of seeds/pollination	No. of seeds/ plant
Alfalfa	1 - 4	600 - 1000
Barley	30 - 40	200 - 600 (greenhouse) 50 - 100 (field)
Canola	1 - 6	260 - 2000
Corn	50 - 500	50 - 600
Dry peas	1 - 4	50 - 200
Soybean	1 - 2	50 - 60
Sunflower	600 - 1000	600 - 1000
Tobacco	100 - 200	10,000 - 1 million
Wheat	30 - 36	200 - 500 (greenhouse) 50 - 60 (field)

Other factors to consider are the availability of acreage and growers, and an infrastructure for handling and processing the material after harvest. Some crops are not widely grown and it may be difficult to find sufficient growers to produce large quantities of a specialty crop. For most protein products, this would not be a problem, but it may be a consideration for production of some industrial enzymes. For pharmaceutical products, use of an obscure crop may have advantages because isolation would be easier to obtain, and the specialized handling required would simplify procedures to keep it out of the food stream.

Seed vs. vegetative tissue

One aspect of choosing a crop species for recominant protein production that has a major impact on all facets of the production system is what plant organ will be harvested as the raw material source. Seeds offer many advantages over vegetative and fleshy fruit tissues. Seeds are natural storage organs intended to store proteins intact until they are degraded as a food source during germination. Leaves and fleshy fruits are dynamic structures where proteins are synthesized and degraded with a high turnover rate. The high processing costs and perishability of vegetative and fruit tissue make them suitable for only small volume, high margin pharmaceuticals. For most other products, grain crops such as corn, soybeans and oil seed rape will provide a more economical production vehicle. Part of the increased cost of production with perishable tissue is due to the necessity that processing must be carefully timed with harvest and the processing facility must be near the crop production site. Another advantage of seeds, especially the larger seeded crops like corn, is that the seed can be fractionated and unneeded parts can be sold off to defray some of the cost of processing and purification (Nikolov, Z., and Hammes, D., this volume). Non-food crops, such as tobacco, contain harmful chemicals (e.g. nicotine) that could contaminate pharmaceutical preparations purified from them and will need to be removed, thus adding to production costs. Food crops, however, present their own regulatory challenges (Emlay, D., this volume).

One of the barriers to plants as a cost effective producer of protein products is low protein yield. Several studies have shown that recombinant protein yield (%TSP) is often higher in seeds than in leaves (Conrad & Fiedler 1998; Giddings *et al.* 2000; Stoger *et al.* 2000). Climatic and stress conditions can dramatically affect expression and accumulation of recombinant proteins. This is true for all plant tissues but is a major problem with vegetative tissues. (Stevens *et al.* 2000) found that transgenic tobacco plants carrying a gene for a mouse IgG had higher amounts of soluble protein (mg/g FW) and IgG when grown at lower temperature (15°C) and high light intensity than they did under high temperature (25°C) and either high or low light intensity. The maturity of the leaf tissue also affected extractable IgG levels, with younger leaves having much higher concentrations of IgG than older, lower leaves. The IgG molecule was proteolytically degraded during leaf senescence. The timing of leaf tissue harvest is therefore critical and will greatly affect product yield and homogeneity. Optimally, the tissue would be processed immediately after harvest, before further degradation and loss of product yield occurs. In contrast, seeds are very stable and can be stored for long periods of time with little or no protein loss (Kusnadi *et al.* 1998; Stoger *et al.* 2000).

Economics of production

Crop factors that have major affects on recombinant protein production costs are: 1) the amount of protein per unit tissue, 2) the cost to produce the crop, and 3) the potential yield of harvestable tissue. Some crops may have the potential to produce large amounts of recombinant protein but are too expensive to produce, while others may be economical to produce but have low yields of both tissue and/or protein (Table 2). As far as sheer volume of recombinant protein produced (kg r- protein/hec), the two leaf crops, alfalfa and tobacco, are potentially far ahead of the grain crops. However the cost to produce the crop may be prohibitive, and as mentioned above, the need for immediate processing at a nearby facility adds further complications, and cost, to purification. Of the grain crops, soybeans and oats are potentially the lowest cost production systems for recombinant protein. The comparison of soybeans and corn is an interesting one, in that by some standards the two crops are very similar, yet if other factors are considered there are clear advantages of one crop over the other. On average one hectare of corn yields 4.5X as much grain as one hectare of soybeans, but it takes 4X as much corn as soybeans to produce 1 kg of recombinant protein. Therefore corn is only slightly better. However, the cost to produce 1 kg of recombinant protein in corn is about 1.7X that of what it costs to produce 1 kg of the same protein in soybeans. From a regulatory standpoint, soybeans have an advantage over corn because they are largely self-pollinated and the risk of contamination through pollen flow is reduced. Soybeans are not without challenges, however. It is one of the more difficult crops to work with genetically, both in terms of plant transformation and genetic improvement through breeding. Patent barriers to the commercial use of this crop in transformation will further limit its use.

Table 2. *Comparison of costs and potential recombinant protein yields from several crops*

Crop	% protein	kg/hec⁺	$/kg⁺	$/hec	kg prot/hec	kg r-protein/hec*	cost of r-protein / kg	kgcrop/kg r-protein
Alfalfa leaves	20	13468	0.11	1481.48	2693.6	269.4	5.50	50.0
Barley	9	3361	0.11	369.75	302.5	30.3	12.22	111.1
Canola	25.5	1347	0.18	237.04	343.4	34.3	6.90	39.2
Corn	10	11313	0.07	746.67	1131.3	113.1	6.60	100.0
Potato tubers	2	36251	0.13	4785.19	725.0	72.5	66.00	500.0
Rice	8	6422	0.13	847.70	513.8	51.4	16.50	125.0
Soybeans	40	2532	0.15	389.93	1012.8	101.3	3.85	25.0
Tomato fruit	1	71829	0.64	45827.16	718.3	71.8	638.00	1000.0
Tobacco leaves	13	33670	4.03	135555.56	4377.1	437.7	309.69	76.9
Wheat	12	2653	0.09	233.48	318.4	31.8	7.33	83.3
Rye	10	1848	0.09	162.67	184.8	18.5	8.80	100.0
Peanuts	25	2875	0.44	1265.19	718.9	71.9	17.60	40.0
Safflower	17	1684	0.31	518.52	286.2	28.6	18.12	58.8
Sunflower	23.5	1411	0.18	248.30	331.5	33.2	7.49	42.6
Oats	15	2357	0.07	155.56	353.5	35.4	4.40	66.7

* Assuming an expression of 10% of soluble protein

⁺ Source: USDA National Agricultural Statistics Service

The selection of a crop species for recombinant protein production may depend on the project and the particular protein being manufactured. One crop may be ideal for certain applications but not for others. Grain crops offer many advantages over vegetative crops and are the most amenable to the production of a wide array of protein products. In seeds, protein accumulation is less affected by time of harvest and environmental conditions, and seeds can be stored for long periods of time without significant protein degradation. Of the grain crops, corn and soybeans have the greatest potential for producing high expression of recombinant proteins and high yields of r-protein per unit area. The transformation of soybeans, however, is less routine than corn, and the anti-nutritional factors in soybean seeds (e.g. soybean trypsin inhibitor and lectin) require that the seed be cooked before consumption. The cooking process would degrade many protein products and would limit the use of soybeans as a direct feed source for administration of vaccines or other pharmaceuticals. Corn does not contain these anti-nutritional factors and can be consumed in raw form by most mammalian species. Of the major crop species, maize is one of the most extensively studied both genetically and physiologically. This wealth of knowledge and germplasm can be exploited to provide an extremely flexible manufacturing tool.

BREEDING TRANSGENIC PLANTS

One would think that the process of breeding a transgenic crop would be much the same as any conventional breeding program. While there are similarities, there are also areas that are handled differently and present unique concerns that are not usually issues for the traditional plant breeder. One major disparity is that the major emphasis is different. For a conventional crop, the main goal is to increase yield and the only concerns are those factors that contribute to yield (e.g. disease and insect resistance, stalk strength, and ear characteristics for corn). For a transgenic crop intended for recombinant protein production, the main emphasis is on obtaining the highest possible expression of the protein of interest, and for a high value protein, such as a pharmaceutical, grain yield (i.e. raw material cost) may not contribute significantly to overall production costs. Lower margin protein products, such as industrial enzymes, where production costs must be kept to a minimum, will need high recombinant protein yield (expression level) and competitive grain yield to make plants a competitive manufacturing vehicle. In addition, there are concerns such as stability of the transgene and transgene expression, genetic make up of the production line (homozygous vs. hemizygous, dosage, etc.), and regulatory hurdles that are rarely an issue in a conventional breeding program.

Stability of transgene expression

Probably the major difference between transgenic plant breeding and conventional plant breeding is that transgenes may, or may not, follow normal Mendelian inheritance, especially if there are multiple copies of the gene or additional copies are added through conventional genetic crossing. The challenge is to minimize the possibility that gene silencing will occur, and then to further cull out any undesirable types through careful selection. Transgene silencing is not a new phenomenon which arose with the development of genetic engineering technology, but occurs as a result of the natural host genome defense system (Matzke *et al.* 2000). This host defense system operates on two fronts. One is active in the nucleus and is associated with *de novo* DNA methylation and the other operates in the cytoplasm and involves sequence-specific RNA degradation. Many excellent reviews of this subject are available from experts in the field, therefore only a brief overview will be presented here (Plant Molecular Biology (2000) 43(2/3); (Matzke & Matzke 1995); (Finnegan & McElroy 1996); (Vaucheret *et al.* 1998a); (Grant 1999); (Kooter *et al.* 1999); (Wolffe & Matzke 1999).

Transgene silencing in plants occurs mainly by two mechanisms 1) transcriptional gene silencing and 2) post-transcriptional gene silencing. Transcriptional gene silencing occurs in the nucleus and is usually associated with methylation of the promoter region (Vaucheret *et al.* 1998a); (Matzke & Matzke 1995); (Kooter *et al.* 1999); (Finnegan & McElroy 1996). It is meiotically inherited (Park *et al.* 1996), but can be reversed slowly over several generations as DNA methylation within the silenced promoter region gradually decreases (Matzke & Matzke 1991). Transcriptional gene silencing occurs most often when multiple copies of a transgene are present with significant homology in the promoter region (Que & Jorgensen 1998). This can occur by the insertion of multiple copies of a transgene, either at one (*cis*) or more (*trans*) loci, by successive transformations with transcripts driven by the same promoter, or through sexual crossing of independent transformation events driven by the same promoter. Single copy insertions of non-native gene sequences can be recognized as foreign by the host genome defense system and lead to methylation of the promoter and silencing (Linn *et al.* 1990); (Meyer *et al.* 1992). The unintentional integration of flanking vector sequences are also recognized as foreign and can target a transgene for silencing (Iglesias *et al.* 1997). Similarly, integration of a transgene into a heterochromatic region of the genome that is normally hypermethylated, disrupts the native chromatin structure and often leads to transcriptional gene silencing (Iglesias *et al.* 1997).

In contrast, post-transcriptional gene silencing occurs mainly in the cytoplasm and is usually associated with methylation of the gene sequence and premature, sequence-specific RNA degradation. This methylation is not heritable and is reset at meiosis (Kunz *et al.* 1996). Post-transcriptional gene silencing often increases during the growing season (Palauqui & Vaucheret 1995) and is often influenced by environmental conditions (Meyer *et al.* 1992). Post-transcriptional gene silencing occurs most when transgenes are driven by strong promoters, such as the strong viral 35S promoter, and more often in haploids and homozygous plants than in hemizygotes (Dehio & Schell 1994); (Elmayan & Vaucheret 1996). The use of a very strong promoter may produce abnormally high transcript levels that trigger a feedback mechanism leading to RNA degradation and methylation of the gene sequence. Alternatively, a higher than normal amount of aberrant RNA may be produced from RNA polymerase errors that may also trigger silencing (Wassenegger *et al.* 1994). Repetitive sequences within the transgene sequence may produce DNA-DNA interactions and abnormal secondary structure leading to malformed mRNA sequences that are targeted for degradation (Vaucheret *et al.* 1998a). Post-transcriptional gene silencing can also cause trans-inactivation of homologous, endogenous gene sequences or other homologous transgenes, a phenomenon called co-suppression (Que & Jorgensen 1998). Again, this occurs most often when the transgene is driven by a very strong promoter.

The two mechanisms of silencing (transcriptional and post-transcriptional) are not mutually exclusive in that both can occur at the same time and to varying degrees. (DeNeve *et al.* 1999) found cases where transgenic *Arabidopsis* plants showing silencing of an antibody gene had methylation in both the promoter and gene sequences. In order for plants to be an attractive vehicle for the production of recombinant proteins, stable, high expression of the transgene is a necessity. Several strategies to avoid the problem of transgene is a necessity. Several strategies to avoid the problem of transgene silencing have been devised and the most successful vectors have no doubt incorporated one or more of these during construction and development. As mentioned above, transgenes can be targeted as 'foreign' invading DNA sequences by the plant genome defense system and silenced, especially if the sequence is from a different organism that has a different isochore, or GC, composition to that of the host species, or if integration occurs in a normally untranscribed region of the chromosome. When at all possible, the transgene sequence should be 'optimized' for the host organism's preferred codon usage, and optimization tables are available for several different crop and model plant species. Controlling the site of integration is more problematic because the most widely used transformation methods, *Agrobacterium*-mediated and particle bombardment, tend to integrate randomly throughout the genome, although *Agrobacterium* transferred DNAs may integrate more often in transcribed regions of the genome. A method for site-specific recombination using the bacteriophage P1 *cre-lox* system has been used experimentally to integrate DNA at *lox* sites in tobacco (Albert *et al.* 1995) but it has not been used commercially. Another method purported to isolate and protect transgenes from the surrounding genomic environment is the incorporation of matrix attachment regions (MARs) into flanking regions of the transgene sequence (Allen *et al.* 2000 and references therein). Using particle bombardment or transfection, constructs with MARs have a lower proportion of non-expressing regenerated plants than when compared to similar constructs without MARs. MARs have also been shown to protect against transcriptional gene silencing in *cis*, especially when multiple copies of the transgene are present (Vaucheret *et al.* 1998b). However, MARs do not prevent transcriptional gene silencing in *trans*, nor post-transcriptional gene silencing (Vaucheret *et al.* 1998b). The choice of a promoter sequence can either protect or predispose a transgene to silencing. As indicated above, very strong promoters, such as the CaMV 35S promoter in leaves, have been shown to make transgenes driven by them more susceptible to silencing by both transcriptional and post-transcriptional mechanisms. The use of native host promoters, either constitutive or tissue specific, can greatly reduce gene silencing and still produce high expressing transformants (ProdiGene, unpublished data).

Transformants with multiple copies of the transgene are more prone to silencing, and measures should be taken to either create single copy transformants or select them out of the pool of transformation events produced. *Agrobacterium*-mediated transformation systems have been shown to mainly produce transformation events with low copy or single-copy integrations of transgene sequences (DeWilde *et al.* 2000); ProdiGene unpublished data), whereas direct gene transfer methods often produce transformants with multiple copies of the transgene (Meyer 1998). In the past, *Agrobacterium* was of limited use as a transformation tool, but today transformation systems have been developed for several of the major crop species including corn and soybeans. A method to ensure only single copy events has been devised for wheat, based on *cre-lox* site-specific recombination (Srivastava *et al.* 1999). The strategy employs a transformation vector composed of a transgene flanked by recombination sites in an inverted orientation. Even if multiple copies of the transgene are integrated, recombination between the outermost sites reduces the integrated molecules to a single copy. And, if more sophisticated methods are unavailable or impractical, molecular methods such as PCR or Southern blotting can be used to screen promising transformation events for those containing single or low copy insertions (DeNeve *et al.* 1998).

Another integration problem that can cause gene silencing in *Agrobacterium*-mediated transformation systems is the incorporation of non-T-DNA sequences into the plant genome. DNA integration most often initiates from the RB and occasionally from the LB, and if it fails to terminate at the opposing border, unwanted vector backbone sequences can be incorporated (Hanson *et al.* 1999; Iglesias *et al.* 1997). These bacterial sequences can not only target a transgene for silencing but can also complicate regulatory approval of a transformed line if it also contains an antibiotic resistance gene. Promising events can be screened by PCR or Southern blotting to identify those that do not contain non-T-DNA flanking sequences. Alternatively, specific vectors can be constructed that will select against events containing segments of DNA from outside the left and right borders. One such vector was designed with a lethal barnase gene inserted downstream of the LB (Hanson *et al.* 1999). Events containing vector sequences beyond the LB die in the callus stage from barnase expression.

The technique has been used successfully in tobacco, tomato and grape.

Strategies for selection of stable high expressing lines

Even when events with single copy, clean integrations are selected, variation for transgene expression is similar to what one would normally attribute to a quantitative trait, i.e. a trait that is typically, but not always, controlled by several genes with small individual effects. This variation occurs among plants from different events (independent gene transfers), but also among plants within an event, which are clonal. The differences in expression between events could be attributed to variations in the chromatin microclimate surrounding the transgene (position effects), which can affect how efficiently the gene is transcribed. Variation between plants within an event is likely due to variations in the allele composition of host genes, which in some way influence expression of the transgene. Once a transformed plant is regenerated from culture, in order to propagate it one must either self pollinate it or cross it to some other germplasm. Unless the host genotype used in the transformation is completely homozygous, either one of those actions will change the genetic composition of the resulting seed and make each progeny unique. The advantage that a transgenic plant breeder has over a conventional breeder selecting for a quantitative trait such as yield, is that the presence or absence of the gene can be easily identified and the gene product can be quantified.

One aspect that transgenic plant breeding has in common with conventional breeding is that in the early stages it is very much a numbers game. The breeder needs to examine all the variation available because things can change from one generation to the next. The favorite event in T1 may not necessarily be the favorite in T2. The most desirable events, and plants within an event, are those that consistently produce progeny that are consistently high expressing over different generations and across different environments. The challenge is to look at enough material to identify that "magic" plant that has both stable and high expression of the transgene. Typically, many events and plants/event of each construct should be produced and examined in the T1. Often there are clear winners and losers, and several events and plants/event are carried forward. Single seed expression levels in T1 seed vary widely but indicate what the potential for expression of the recombinant protein might be (Fig.1). For corn, T0 plants are typically outcrossed to an elite inbred, therefore the resulting T1 seed is segregating for hemizygous and null kernels. Promising T1 lines are then planted in field nurseries and are self-pollinated. Selfing accomplishes two objectives: 1) it allows the breeder to identify and cull out low expressing lines which may be undergoing gene silencing in the homozygous state, 2) it allows the breeder to determine gene action for the protein. If expression does not increase in selfed progeny, then gene action is not additive and there may be no advantage to homozygous over hemizygous lines. This knowledge will become important when choosing a final production strategy. In T2 and higher generations, recombinant protein expression is quantified in bulk seed samples, which is a better estimate of what production samples will achieve. It is common for the range of expression of bulk seed samples to decrease, as the contribution of high expressing single seeds is buffered by other lower expressing seeds in the bulk (Fig. 1). In the T3 generation, we begin backcrossing to elite inbred germplasm, and since these are now hemizygous populations of seeds it is expected that expression levels will decrease slightly from the selfed progeny in the previous generation. By this stage, the non-expressors have been eliminated, and in T4 and further generations the overall mean of the population begins to increase rapidly as the selections are narrowed down and the genetic background of the lines improves.

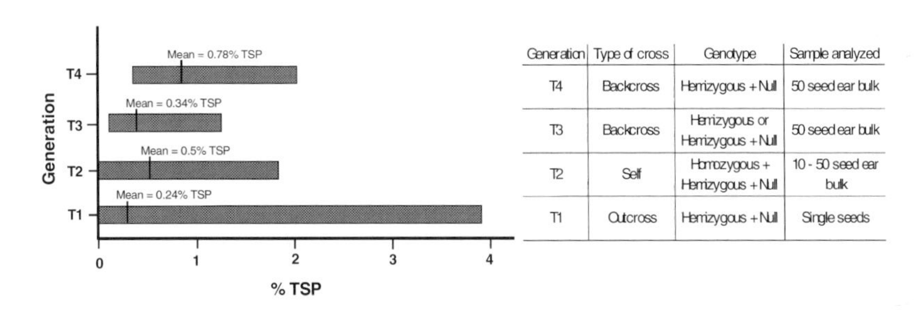

Generation	Type of cross	Genotype	Sample analyzed
T4	Backcross	Hemizygous + Null	50 seed ear bulk
T3	Backcross	Hemizygous or Hemizygous + Null	50 seed ear bulk
T2	Self	Homozygous + Hemizygous + Null	10 - 50 seed ear bulk
T1	Outcross	Hemizygous + Null	Single seeds

Figure 1. *Progress from Selection*

Throughout this process, it is important to note any unusual characteristics of the transformed lines, such as male sterility, plant health, or environmental interactions. Often there are no overall affects of the transgene that would distinguish a transformed plant from a non-transformed plant, but certain environments may consistently produce higher expression of the transgene than others. These should be noted because it may be important that production lots be grown in environments known to maximize expression.

Effect of genetic background

To date, little consideration has been given to the effect of different genetic backgrounds on transgene expression because the transgenic products that are commercially available are more targeted toward achieving a phenotype (i.e. herbicide tolerance or insect resistance) than with protein production. When the goal is to produce as much recombinant protein in as little tissue as possible, every contributing factor (vector construction, transformation method, genetic background, growing environment, etc.) that influences transgene expression should be optimized. Pairing the right transgene with the right genetic background can easily double or triple transgene expression. There are two components that contribute to increasing recombinant protein yields 1) maximize the yield of harvested tissue (e.g. leaves or seeds) and 2) maximize the yield of recombinant protein per unit tissue. Depending on the value of the protein product, factor 2 may be more important than factor 1. The yield of r-protein per unit tissue is influenced by the amount of extractable protein per unit tissue and transgene expression level (% TSP, percent of total soluble protein). The genetic background that a transgene is expressed in can affect all of these factors (Table 3). A single event of transgenic corn expressing a small protein product was crossed into five different genetic backgrounds. In backgrounds 1 and 2, the expression level of the transgene was the same (0.69 % TSP), but background 1 has almost twice the amount of total extractable protein as background 2, and thus has a higher yield of r-protein per seed. If seed yield is considered, to obtain an equivalent amount of r-protein/acre, background 2 would need to achieve a yield of 270 Bu/acre, whereas for background 1 a yield of 150 Bu/acre would be sufficient. Yields of 150 Bu/acre are common in most commercial corn growing areas, however yields of 270 Bu/acre or more would require the best hybrid germplasm under the best growing conditions. This illustrates that more benefit can be realized from increasing the amount of extractable protein per unit tissue than by increasing yield. Increasing expression level (% TSP) (backgrounds 3, 4, and 5) will also increase the yield of r-protein/acre, and if high expression is combined with high extractable protein and high seed yield, that is of course the best combination of all (Table 3).

Table 3. *Comparison of yield of recombinant protein per acre in different genetic backgrounds for a single event of transgenic corn.*

Background	% TSP	mg r-protein/ kg seed	g r-protein / acre	
			150 Bu/acre[a]	270 Bu/acre[b]
1	0.69	97.87	373.7	672.6
2	0.69	55.90	213.4	384.2
3	0.77	86.40	329.9	593.8
4	0.83	118.10	450.9	811.7
5	0.96	149.60	571.2	1028.2

[a] 9427 kg/hectare

[b] 16970 kg/hectare

The gene product of some transgenes can have deleterious effects on plant health and growth characteristics, and for these products finding the right germplasm background to express it in can make the difference between success or failure of the project. For example, one enzyme product causes severe affects on plant health, including leaf tip burning, poor germination, barren plants, and premature death. However, we were able to select healthier individuals from the population. One solution to the problem would have been to express the gene under a promoter that did not function in leaf or stem tissue. Another approach would be to cross it into other germplasm backgrounds in the hope of finding one that would keep the enzyme largely inactive until it was extracted. By applying both strategies, we were successful and now have lines that show no ill effects on plant health, double the germination and increase the seed yield to over 10-fold (Table 4).

Table 4. *Effect of genetic background on plant growth and transgene expression for a transgene with deleterious effects.*

Background	Germination (%)	Plant health	Yield (Bu/acre)	mg r-protein/ kg seed
1	40	Heavily damaged	8	33
2	70 - 80	No damage	95	76

Many plant species contain compounds that can complicate purification and downstream processing of certain products (e.g. nicotine in tobacco, anti-nutritional factors in soybean, corn trypsin inhibitor in maize, etc.). Expression of transgenes in backgrounds that lack these compounds would reduce processing costs and increase purification yields.

Production Strategies

Seed from transgenic lines developed for protein production (i.e. pharmaceutical or industrial) is generally not intended for commercial sale. The crop is grown purely for the protein product it contains, is owned solely by the company from which it originated, and is not sold to growers as seed. This gives the production team much more flexibility in how the crop is produced, and reduces the time required from initiation of the project to the commercial production phase. For example, if the protein product is being produced in corn, since there are no guarantees made on the quality of the crop, there is no need to develop a finished, fully tested hybrid before it is moved into production. The final production line may indeed be a finished hybrid, but production can begin before that is completed. In the fast-paced world of drug development, this is a real advantage. A company can therefore follow two production strategies. The short-term strategy is to identify a stable, high expressing line from which a predictable amount of the protein product can be produced, regardless of the fact that it may not be the most efficient or economical. The long-term goal should be to identify the best construct, the highest expressing event, in a genetic background that provides the highest yield of both the protein product and harvested tissue per unit area.

The discovery phase required to achieve the long-term goal may take years but the payoff can be enormous as far as lowering the cost of downstream processing (Nikolov, Z., and Hammes, D., this volume). The options available to the plant breeder will depend on the crop, and somewhat on the value, and volume, of the protein product being produced. For a self-pollinated crop like soybeans, most production lines will either be a single or mixed event population (early phase) or an inbred (finished line), and only for a very high volume product would a hybrid be considered. However, for a cross-pollinated crop like corn, either a mixed population or a finished hybrid would be the most likely choices. As mentioned above, another factor to consider is the gene action of the transgene. Is there a dosage effect? Are homozygous plants higher expressing than hemizygous? If not, then the extra time required to develop a homozygous production line may not be justified. Conversely, if a double dose of the transgene does improve expression, a homozygous production line will offer an advantage, and one should also investigate whether additional doses would further increase expression. Pyramiding genes could be accomplished in a number of ways, either by 1) crossing two different events together, 2) crossing two different constructs of the gene together, each under the control of a different promoter, or 3) transforming with a multiple copy (multi-plant transformation unit) construct. It should be kept in mind that any of these attempts could be subject to gene silencing. Two genes under the same promoter may be subject to transcriptional gene silencing, whereas two genes under different promoters may be silenced post-transcriptionally. However, there are cases where pyramiding transgenes has been successful. At ProdiGene we have produced transgenic corn lines homozygous for two different GUS events that have double the expression of a single event homozygote (ProdiGene, unpublished data). For ease of regulatory approval, a single event production line is favored, but if the increase in transgene expression is sizeable enough, the extra cost to move two events through the regulatory process may be worthwhile. This is especially true for industrial enzymes for which the regulatory approval process is less rigorous than for pharmaceuticals.

Certain pharmaceuticals, or other proteins that may pose a health or environmental risk will require specialized production lines or more elaborate containment conditions to prevent unwanted exposure or contamination of the food stream. These products may be best produced in a self-pollinated crop. For cross-pollinated crops, male-sterile lines could be developed or the transgenic line could be detasseled and used only as a female. Some companies have developed postharvest expression systems that utilize inducible promoters that are turned on in response to wound-induction or mechanical stress (Cramer *et al.* 1999), or during seed germination (malted barley) (Rodriguez 1999). These promoters are not normally activated during plant growth, but can be induced postharvest under controlled conditions. Another molecular method of transgene isolation has been proposed which would utilize two genes, a blocking sequence and a recovering sequence (Kuvshinov *et al.* 2001). The blocking sequence blocks a molecular or physiological function of the host plant rendering it either unable to germinate or incapable of reproducing. The recovering sequence is inducible by either a chemical or physical treatment of the plants and does not function under natural conditions. Plants carrying the introduced sequences will only germinate and/or reproduce if treated, thus activating the recovering sequence. A system that would be practical under field conditions using this concept has not yet been devised.

The flexibility, stability, and efficiency in both time and cost that plant production systems offer have advantages over all other protein production methods. As the technology advances and the infrastructure necessary to isolate these products from food crops becomes established and routine, the benefits in low cost pharmaceuticals and safer, more efficient industrial manufacturing methods will be appreciated.

REFERENCES

Albert H, Dale E, Lee E, Ow D. (1995) Site-specific integration of DNA into wild-type and mutant lox sites placed in the plant genome. Plant Journal 7: 649-659

Allen G, Spiker S, Thompson W. (2000) Use of matrix attachment regions (MARS) to minimize transgene silencing. Plant Mol Biol 43: 361-376

Comai L. (2000) Genetic and epigenetic interactions in allopolyploid plants. Plant Mol Biol 43: 387-399

Conrad U, Fiedler U. (1998) Compartment-specific accumulation of recombinant immunoglobulins in plant cells:an essential tool for antibody production and immunomodulation of physiological functions and pathogen activity. Plant Mol Biol 38: 101-109

Cramer C, Boothe J, OIshi K. (1999) Transgenic plants for therapeutic proteins: linking upstream and downstream strategies. In J Hammond, P McGarvey, V Yusibov, eds Plant Biotechnology: New products and applications. Springer-Verlag, Berlin, pp 95-118

Dehio C, Schell J. (1994) Identification of plant genetic loci involved in a posttranscriptional mechanism for meiotically reversible transgene silencing. Proceedings of the National Academy of Sciences USA 91: 5538-5542

DeNeve M, DeBuck S, DeWilde C, Van Houdt H, Strobbe I, Jacobs A, Van Montagu M, Depicker A. (1999) Gene silencing results in instability of antibody production in transgenic plants. Mol Gen Genet 260: 582-592

DeNeve M, Van Houdt H, Bruyns A-M, Van Montagu M, Depicker A. (1998) Screening for transgenic lines with stable and suitable accumulation levels of a heterologous protein. Methods in Biotechnology 3: 203-227

DeWilde C, Van Houdt H, DeBuck S, Angenon G, DeJaeger G, Depicker A. (2000) Plants as bioreactors for protein production: avoiding the problem of transgene silencing. Plant Mol Biol 43: 347-359

Elmayan T, Vaucheret H. (1996) Expression of single copies of a strongly expressed 35S transgene can be silenced post-transcriptionally. Plant Journal 9: 787-797

Finnegan J, McElroy D. (1996) Transgene stability. In M Owen, J Pen, eds Transgenic plants: A production system for industrial and pharmaceutical proteins. J Wiley & Sons, New York, pp 169-186

Giddings G, Allison G, Brooks D, Carter A. (2000) Transgenic plants as factories for biopharmaceuticals. Nature Biotechnology 18: 1151-1155

Grant S. (1999) Dissecting the mechanisms of posttranscriptional gene silencing: divide and conquer. Cell 96: 303-306

Hanson B, Engler D, Moy Y, Newman B, Ralston E, Gutterson N. (1999) A simple method to enrich an Agrobacterium-transformed population for plants containing only T-DNA sequences. Plant Journal 19: 727-734

Hovenkamp-Hermelink J, Jacobsen E, Pijnacker L, deVries J, Witholt B, Feenstra W. (1988) Cytological studies on adventitious shoots and minitubers of a monoploid potato clone. Euphytica 39: 213-219

Iglesias V, Moscone E, Papp I, Neuhuber F, Michalowski S, Phelan T, Spiker S, Matzke M, Matzke A. (1997) Molecular and cytogenetic analyses of stably and unstably expressed transgene loci in tobacco. Plant Cell 9: 1251-1264

Kooter J, Matzke M, Meyer P. (1999) Listening to the silent genes: Transgene silencing, gene regulation and pathogen control. Trends in Plant Science 4: 340-347

Kunz C, Schob H, Stam M, Kooter J, Meins F. (1996) Developmentally regulated silencing and reactivation of tobacco chitinase transgene expression. Plant Journal 10: 437-450

Kusnadi A, Hood E, Witcher D, Howard J, Nikolov Z. (1998) Production and purification of two recombinant proteins from transgenic corn. Biotechnol Prog 14: 149-155

Kuvshinov V, Koivu K, Kanerva A, Pehu E. (2001) Molecular control of transgene escape from genetically modified plants. Plant Science 160: 517-522

Linn F, Heidmann I, Saedler H, Meyer P. (1990) Epigenetic changes in the expression of the maize A1 gene in Petunia hybrida: role of numbers of integrated gene copies and state of methylation. Mol Gen Genet 222: 329-336

Matzke M, Matzke A. (1991) Differential inactivation and methylation of a transgene in plants by two suppressor loci containing homologous sequences. Plant Mol Biol 16: 821-830

Matzke M, Matzke A. (1995) How and why do plants inactivate homologous transgenes? Plant Physiology 107: 679-685

Matzke M, Mette M, Matzke A. (2000) Transgene silencing by the host genome defense: implications for the evolution of epigenetic control mechanisms in plants and vertebrates. Plant Mol Biol 43: 401-415

Meyer P. (1998) Stabilities and instabilities in transgene expression. In K Lindsey, ed Transgenic Plant Research. Harwood Academic Publishers, Amsterdam, pp 263-275

Meyer P, Linn F, Heidmann I. (1992) Endogenous and environmental factors influence 35S promoter methylation of a maize A1 gene construct in transgenic petunia and its colour phenotype. Mol Gen Genet 231: 352

Mittelsten Scheid O, Jakovleva L, Afsar K, Maluszynska J, Paszkowski J. (1996) A change of ploidy can modify epigenetic silencing. Proceedings of the National Academy of Sciences USA 93: 7114-7119

Palauqui J-C, Vaucheret H. (1995) Field trial analysis of nitrate reductase cosuppression: a comparative study of 38 combinations of transgene loci. Plant Mol Biol 29: 149-159

Park Y-D, Papp I, Moscone E, Iglesias V, Vaucheret H, Matzke A, Matzke M. (1996) Gene silencing mediated by promoter homology occurs at the level of transcription and results in meiotically heritable alterations in methylation and gene activity. Plant Journal 9: 183-194

Pawlowski W, Torbert K, Rines H, Somers D. (1998) Irregular patterns of transgene silencing in allohexaploid oat. Plant Mol Biol 38: 597-607

Que Q, Jorgensen R. (1998) Homology-based control of gene expression patterns in transgenic petunia flowers. Developmental Genetics 22: 100-109

Rodriguez R. (1999) Process for protein production in plants. United States Patents 5,888,789, 5,889,189, and 5,994,628

Srivastava V, Anderson O, Ow D. (1999) Single-copy transgenic wheat generated through the resolution of complex integration patterns. Proceedings of the National Academy of Sciences USA 96: 11117-11121

Stevens L, Stoopen G, Elbers I, Molthoff J, Bakker H, Lommen A, Bosch D, Jordi W. (2000) Effect of climate conditions and plant developmental stage on the stability of antibodies expressed in transgenic tobacco. Plant Physiology 124: 173-182

Stoger E, Vaquero C, Torres E, Sack M, Nicholson L, Drossard J, Williams S, Keen D, Perrin Y, Christou P, Fischer R. (2000) Cereal crops as viable production and storage systems for pharmaceutical scFv antibodies. Plant Mol Biol 42: 583-590

Vaucheret H, Beclin C, Elmayan T, Feurerbach F, Gordon C, Morel J-B, Mourrain P, Palauqui J-C, Vernhettes S. (1998a) Transgene-induced gene silencing in plants. Plant Journal 16: 651-659

Vaucheret H, Elmayan T, Thierry D, van der Geest A, Hall T, Conner A, Mlynarova L, Nap J-P. (1998b) Flank matrix attachment regions (MARs) from chicken, bean, yeast or tobacco do not prevent homology-dependent trans-silencing in transgenic tobacco plants. Mol Gen Genet 259: 388-392

Wassenegger M, Heimes S, Riedel L, Sanger H. (1994) RNA-directed de novo methylation of genomic sequences in plants. Cell 76: 567-576

Wolffe A, Matzke M. (1999) Epigenetics: Regulation through repression. Science 286: 481-486

Wolters A-M, Visser R. (2000) Gene silencing in potato: allelic differences and effect of ploidy. Plant Mol Biol 43: 377-386

PRODUCTION OF RECOMBINANT PROTEINS FROM TRANSGENIC CROPS

Zivko Nikolov and Daniel Hammes*

ProdiGene
101 Gateway Boulevard
College Station, TX 77845

*Quality Traders, Inc.
11922 Oak Creek Parkway
Huntley, IL 60142

INTRODUCTION

Plants are low cost producers of protein, highly cultivated throughout the world with well-developed infrastructures for processing and handling. In addition, plants produce renewable products that utilize sunlight, soil and air and provide a good "delivery" mechanism, especially for products that can be utilized in food, feed or biotechnology applications. As should be abundantly clear from the other contributions to this book, plants offer a viable option for producing recombinant protein products.

In a broader sense, protein production from transgenic plants consists of two major components, transgenic crop production and bioprocessing. Each of these components requires special considerations. For instance, crop production requires field isolation of the transgenic crop as well as methods for identity preservation (IP) and containment. Additionally, special regulatory concerns must be addressed and safeguards for quality control must be put in place. Bioprocessing, on the other hand, may include some type of crop grinding, extraction of the target protein from the ground material and purification. The cost of the final protein product is a composite of the cost of these two components and depends largely on the type of crop chosen and the final application of the protein produced. This chapter will address specific production and bioprocessing features of transgenic plant systems that utilize environmentally safe and economical practices to manufacture a variety of protein products.

E.E. Hood and J.A. Howard (eds.), Plants as Factories for Protein Production, 159–174.

PLANT PRODUCTION SYTSEMS

The two most common systems today for production of recombinant protein products in plants are tobacco and corn. However, other crops are under consideration for commercial protein production including alfalfa, barley, canola, rice, and safflower. The bulk of our discussion here will concentrate on corn, tobacco, and alfalfa. However, other crops are under consideration for commercial protein production including alfalfa, barley, canola, rice and safflower. In addition, edible plants and plant organs such as beets, peas, beans, carrots, tomatoes and potatoes may also provide viable production systems even though their potential has yet to be demonstrated (Theisen, 1999). While many of these crops may have unique advantages, all potential plant production systems must eventually be evaluated in terms of productivity and cost.

As shown in Table 1, a number of factors should be considered when analyzing plant production systems. Not only total crop yield, but also the total protein yield from each crop should be taken into account. This of course must be balanced against the cost of growing the crop as well as the costs of downstream processing of a recombinant protein from different plant tissues. To simplify the comparison of different plant systems, the transgenic crop cost listed in Table 1 is assumed to be twice that of commercial crops.

Recombinant protein production in alfalfa is the least expensive ($11/kg) among the crops listed in Table 1. Crop protein yields per acre for alfalfa are 5 to 10 fold greater than other crops while its growing cost is only 2 to 6 fold greater. Canola and corn compare favorably with alfalfa although it is interesting to note that the similarity in cost between the two crops stems from different sources. Corn is 3 times more expensive to grow than canola, but its yield per acre is almost 8 times higher than that of canola, resulting in nearly 3 times more protein per acre. By contrast, tobacco is the most expensive crop to grow ($7,320/acre) and has the lowest protein yield (118 kg/acre).

The last two columns in Table 1 compare recombinant protein weight concentrations and their anticipated concentrations in the aqueous extract. Tobacco and alfalfa concentrations are expressed per fresh weight while those of seeds are expressed per dry weight. The assumption for estimating the extract concentration is that chopped fresh tissue is extracted at a biomass to buffer ratio of 1:2 whereas that of ground seed is extracted at a ratio of 1:4. These numbers are useful for estimating the comparative cost of downstream processing. By using the assumption that recombinant protein will be 10% of the total protein, the predicted recombinant concentrations in the tissue and extract are proportional to the crop protein concentration. These estimates are much greater than most reported numbers to date. For example, accumulation between 0.01 and 0.4% fresh weight (Garger *et al.*, 2000) have been measured in tobacco and 0.4% dry weight in corn (unpublished data). However, current trends in improved technology suggest that expression levels such as those in Table 1 should be possible in the future.

Related to the production economics of any potential crop plant is the decision to target expression to harvestable tissues and organs. Targeted expression may be preferred for several reasons. First, using tissue or organ-specific expression allows more efficient utilization of plant energy resources compared to constitutive expression, thereby reducing potential interference with basic metabolism. Second, expression from tissue or organ-specific promoters is less likely to affect plant growth and more likely to yield greater accumulation levels of heterologous proteins. Finally, if the targeted tissue is a harvestable storage organ, bioprocessing cost and regulatory requirements may be less complex.

The green tissues of tobacco and alfalfa, although harvestable, do not have the same advantages as roots, tubers and seed. Leaves are not considered to be storage organs and large accumulation of protein in leaves may affect plant health. Leaf-specific promoters are not as specific as promoters for storage organs, leading to the possibility of protein accumulation in other parts of the plant and requiring the same regulatory procedures for biomass containment and disposal as with constitutive expression.

Table 1. *Cost of Recombinant Protein Production from Different Transgenic Crops*

Crop	Crop Yield (kg /acre)	Protein Conc. (% w/w)	Crop Protein Yield (kg /acre)	Rec. Protein Yield[2] (kg/acre)	Transgenic Crop Cost[1] ($/acre)	Rec. Protein Cost ($/kg)	Rec. Protein Conc. (% wt)	Rec. Protein in the extract (g/L)[3]
Green Tissue								
Alfafa leaves	5455	20	1091	109	1200	11	2.0	6.7
Tobacco leaves	909	13	118	12	7320	610	1.3	4.3
Grain								
Barley	1364	9	123	12	300	25	0.9	3.0
Canola	545	25	136	14	192	14	2.6	8.7
Corn	4091	10	409	41	604	15	1.0	3.3
Rice	2636	8	211	21	688	33	0.8	2.7
Safflower	682	17	116	12	420	35	1.8	6.0

[1] IP cost assumed to be twice the price of commercial crops
[2] Assumed 10% total protein expression
[3] Leaves extracted with buffer at 1:2 ratio; dry seed extracted at 1:4 ratio

PRODUCTION SYSTEM REQUIREMENTS

Identity preservation systems for value-enhanced grains such as corn have been in existence for many years. Seed corn is the single largest example, although other value-enhanced crops (VEC's) are produced in significant volumes (Table 2). Identity preservation systems such as food grade corn, white, waxy and high oil corn can vary in tolerance for contamination of non-target grain, but concern for "out-crossing" to commodity corn is generally not an issue. Requirements for containment and quality control will be significantly stricter for many recombinant protein products than those for VECs. For this reason, the industry will need new containment systems, and perhaps even new terminology, to distinguish between identity preservation systems for food and feed versus the systems for biopharmaceutical and industrial proteins.

Regardless of the plant system utilized, all will require the following:

- Isolation from other crops if out-crossing or co-mingling can occur.
- Identity preservation beginning with production of planting seed through final product formulation. This includes containment of the seeds, leaves and other plant parts, especially between harvest and processing, including residue management and volunteer plant control during subsequent crops.
- Regulatory compliance including USDA permits for planting and movement of harvested crop.
- Quality control and quality assurance programs in conjunction with Standard Operating Procedures (SOP's) to ensure final product quality and regulatory compliance.

Table 2. *Value-enhanced corn production estimates*

Corn Product	Estimated 1998 Acreage (000)	Estimated 1999 Acreage (000)	Projected 2000 Acreage (000)	Projected 2000 Producer Premiums (per metric ton)	Projected 2000 Premiums (per bushel)	Growth Projection
White	725	1,100	900	$9.84-$13.78	$0.25-$0.35	Decline
Waxy	500	550	550	$5.91-$11.81	$0.15-$0.30	Stable
Hard endosperm/food grade	800-1,200	800-1,200	1,000-1,2000	$3.94-$5.91	$0.10-$0.15	Low
High oil	900	1,000	1,100	$5.91-$9.84	$0.15-$0.25	Medium
Nutritionally enhanced	140	200	225	$5.91-$9.84	$0.15-$0.25	High
High amylose	30-40	40-50	40-50	$43.31+	$1.10+	Stable
Total	3,095-3,505	3,690-4,100	3,815-4,025			
% of U.S. harvested acreage	4.3-4.6%	5.2%-5.8%	5.3%-5.6%			

Source: U.S. Grains Council

Isolation

Isolation requirements will depend upon whether the crop is self-pollinated or capable of cross-pollinating adjacent crops. For self-pollinated crops such as tobacco, isolation distance is based on a reasonable distance to ensure the crop is not mixed at harvest (USDA/APHIS, 2001). Because corn is subject to out-crossing, the USDA requires a minimum of 660 feet of isolation for corn grown under "notification". Products targeted for the industrial protein markets (non-food and feed products for markets such as industrial enzymes or research chemicals) generally fall into this category. This distance is considered reasonable to minimize out-crossing from pollen carrying the transgene that could cross-pollinate commodity corn and result in the production of the recombinant protein in a food or feed crop. The isolation distance is doubled to 1320 feet for pharmaceuticals expressed in corn for either animal or human applications and grown under a USDA "permit" (Hammond, 2000; White, 2000).

The above requirements present significant challenges for producing a transgenic corn crop in major corn growing areas due to the difficulty of finding fields with the required distance from other corn. One potential option is to produce the recombinant corn in areas of the U.S. that have limited corn production, recognizing that costs of production may be higher. Other options may include de-tasseling or male sterility to control pollen, or temporal isolations which rely on differences in the pollination timing between commodity corn and the transgenic line.

Identity Preservation and Containment

Identity preservation and containment are important issues for quality control of the target protein product and in preventing contamination of the commodity grown for food or feed. Production of planting seed should be at the highest level of quality control. This practice is already common in the seed industry for the production of "foundation" seed. Planter clean out is necessary prior to planting the recombinant crop to ensure purity in addition to clean-out after planting to prevent contamination of commodities with the recombinant crop. This is easily accomplished with most planters (Hanna & Greenless, 2000). A much more difficult aspect is the cleaning of harvesting equipment, especially with grain crops such as corn. For small volumes, harvesting ears by hand or by machine in a manner similar to seed corn harvesting may be an option. Dedicated combines or those retrofitted for quick and thorough clean out may be needed as volumes increase. Ultimately, a feed and food tolerance allowing de-regulation of the crop may be desired for large volume products determined to be safe for food and feed consumption.

The same level of containment and clean out is required during drying, storage and transportation of the crop. Systems that validate these steps will likely be required for any crops that do not have the required food or feed tolerances. For example, one company utilizes customized equipment for transportation of recombinant tobacco to assure no plant material is introduced into the environment en-route to the processing plant. These and other systems will be part of an overall quality control and quality assurance program that is based on the required purity of the final product and the need for absolute containment of the target crop.

Once the crop reaches a processing plant, quality and containment issues can generally be addressed through conventional methods. However, if contract processing is utilized, the challenge will be greater than if a dedicated and company-controlled facility is used. Processing strategies will be dependent on product specifications and customer requirements such as GMPs (Good Manufacturing Practices).

Regulatory Compliance

USDA permits govern the planting of any recombinant crop prior to issuance of a feed or food tolerance. This agency also controls the movement of viable seeds that contain the protein up to the point of processing into non-viable fractions. For plant-based animal vaccines, additional movement permits are required for the evaluation of an animal biologic that is regulated by the Veterinary Biologics division of the USDA (Henderson, 2000). A system that monitors the required permits and assures that movement is accompanied by the required documentation will need to be part of the overall production plan.

Protein products for human and animal pharmaceutical application are regulated by the U.S. Food and Drug Administration (FDA). Compliance with FDA regulations is expected to start with the first step in the manufacturing plant. All phases of the manufacturing (grinding, extraction, purification, and formulation) of pharmaceutical products must be conducted in strict compliance with the regulations and guidance currently used for other pharmaceutical proteins.

In the end, each plant-based production system has to be designed to best address a particular crop, protein product and associated containment requirements. Differences in production systems will likely be as diverse as the companies designing those systems and the products being produced.

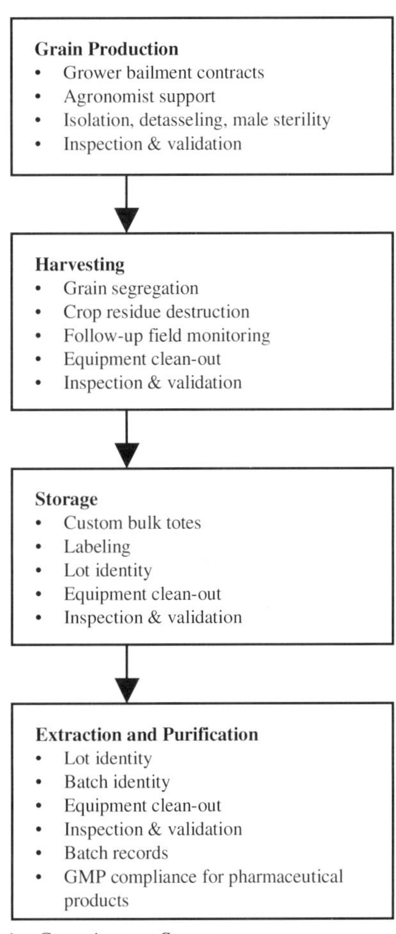

Figure 1. *ProdiGene Identity Containment System*

Figure 1 depicts major elements of a generalized containment system currently utilized by ProdiGene. Under this system, containment procedures are in place for each step of production. SOPs guide the containment with validation built into each critical control point. Different products require different levels of containment, ranging from standard identity preservation for products de-regulated by the USDA, to maximum containment of all plant material in greenhouses. For further details on this subject refer to the regulatory chapter of this book.

Quality Control and Quality Assurance

Any successful production system must have well designed quality control procedures that can be validated and that are relatively easy to administer and follow. Each crucial point in the production system needs to be addressed. Expectations of the grower must be communicated and clearly stated in the contract or growers guide provided by the company responsible for production. Quick and accurate analytical tests will be required to monitor and assure quality throughout processing of the final product. SOPs can be utilized to identify the critical steps of a procedure to third parties who are involved. Validation that these procedures have been followed needs to be built into the system to assure quality and compliance.

BIOPROCESSING OF TRANSGENIC CROPS

Typically, bioprocessing of grain crops would start with the grinding step in the manufacturing facility, whereas bioprocessing of green-tissue crops really begins in the field and continues through the extraction and purification of heterologous proteins. Bioprocessing of different crops presents different challenges in terms of handling, extraction and purification. For simplicity, we will limit our discussion to green tissues, represented by tobacco and alfalfa, and grain, represented by corn.

As an overview, we have diagramed bioprocessing in general terms for green tissues and for grain (Figure 2). Once harvested, green biomass must be processed quickly unless there are compelling reasons to dry and store the plant material. Chopping, homogenizing and pressing the harvested plant tissue results in a "green" juice. The pressed biomass is then discarded and the juice processed further before protein purification can take place. By contrast, grain can be stored for prolonged periods before grinding and protein extraction with aqueous buffer. Centrifugation results in a crude protein extract ready for purification, as well as grain solids that can be processed into salable by-products. In the following section, we discuss these processes in more detail.

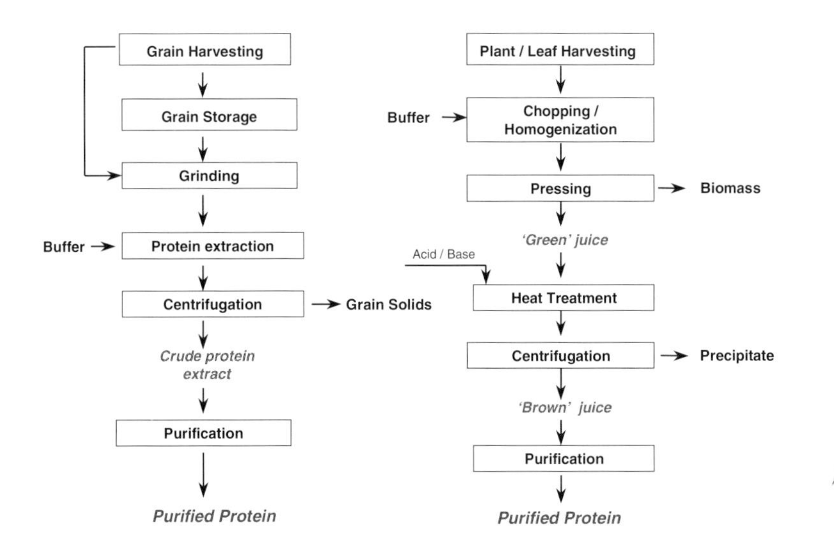

Figure 2. *Bioprocessing of grain and fresh plant tissue for protein production*

Tobacco and alfalfa

Because tobacco is relatively easy to transform and one of the fastest crops to regenerate, it is often used to test the feasibility of producing selected recombinant proteins (Fischer *et al.*, 1999). Alfalfa, on the other hand, is a relative newcomer to the field of plant biotechnology with insufficient published data to date to prove itscommercial value. Nevertheless, alfalfa will likely prove to be a legitimate crop for protein production.

As shown in Table 1, tobacco and alfalfa both yield tens of tons of biomass per acre assuming three to five harvests per year (Cramer *et al.*, 2000; Vezina *et al.*, this volume). Expression of recombinant proteins is generally targeted to leaves since the seed does not represent a significant fraction of the whole plant for the recovery of large quantities of protein. One of the principle challenges facing protein production in crops such as tobacco and alfalfa is the instability of the green biomass. Leaves and green tissue generally cannot be stored after harvest in the same way as seed and tubers. Freezing and lyophilization are possible but may be too costly for the production of all but a few pharmaceutical proteins.

A more likely alternative is to dry the green tissue. Transgenic tobacco leaves expressing ER-retained single chain antibody fragment scFvs were dried and stored for more than three weeks without losses of antigen-binding activity or specificity (Fiedler *et al.*, 1997). Vezina *et al.*, (this volume) reported that mAb (C5-1) was stable for 12 weeks in alfalfa hay dried at room temperature. However, neither study commented on whether the amount of the extractable antibody remained the same after drying.

If the protein of interest is not stable in the leaf during senescence and drying, other alternatives will need to be explored. For instance, growing the crop near the processing plant allows considerable flexibility in matching downstream processing with the harvesting season. Large Scale Biology, Inc. has developed a process whereby tobacco leaves are cut, collected in a specialized stainless-steel trailer and transported to the bioprocessing plant (Barry Holtz, personal communication). The high alkaloid content of tobacco apparently allows harvesting and transport without significant recombinant protein losses. Yet another approach is to macerate and squeeze the juice on the field under aseptic conditions and to stabilize the juice by reducing microbial activity during transport to the processing facility (Austin *et al.*, 1994). In any case, the scale and production economics will dictate the choice of pre-processing of the harvested tissue.

Green tissues also present challenges with regard to purification of heterologous proteins. Typically, grinding and/or homogenization of tissues releases plant pigments such as chlorophyll and thylakoid membranes that need to be removed prior to chromatography. Furthermore, phenolic compounds and other extract components may interact detrimentally with the target protein, thus reducing the recovery yield. (Gegenheimer, 1990; Loomis, 1974; Sumere *et al.*, 1975). Finally, secondary metabolites such as neonicotine (anabasine) and nicotine must usually be removed early in the purification process (Theisen, 1999). Vacuum infiltration of green leaves with buffer, followed by mild centrifugation has been reported for the purification of α-galactosidase A (Turpen, 1999). However, it is not clear if this method can handle high throughputs of biomass (Doran, 2000).

A final consideration in the processing of green tissue is the disposal of biomass. This could potentially be problematic for large volumes, especially when no byproduct revenues are possible as in the case of tobacco. In this, alfalfa may have a distinct advantage if extracted biomass can be used as feed.

Corn

As shown in Table 1, corn has much higher crop yields than tobacco at a much lower cost. By contrast, corn and alfalfa are both quite comparable in terms of cost and yield. However, heterologous protein expression in corn is generally targeted to grain, a natural storage organ requiring no novel harvest or transport technologies. Furthermore, by using tissue specific promoters, recombinant protein accumulation can be targeted to the endosperm or germ, providing an opportunity to pre-fractionate the grain and to reduce the amount of transgenic biomass to be extracted (Figure 3).

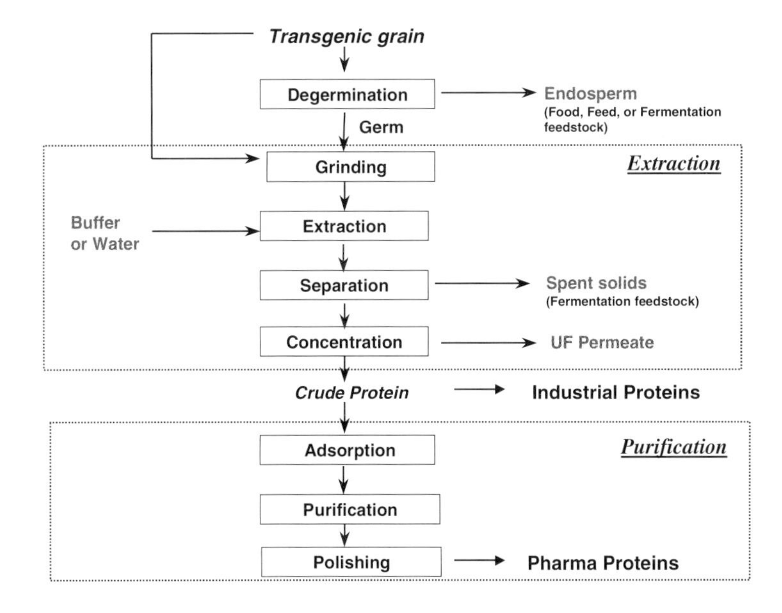

Figure 3. *Bioprocess Products and Byproducts*

Apart from the production system requirements mentioned above for identity preservation and containment, the existing agricultural infrastructure is sufficient in most cases for harvest and transport of genetically modified corn. Furthermore, grain can be stored for considerable time periods with only minimal protein loss. For instance, recombinant β-glucuronidase and avidin expressed in corn were stable for months when stored at room temperature (Kusnadi *et al.*, 1998), and β-glucuronidase for a year when stored at 10°C (Yildirim *et al.*, 2001). This allows for considerable flexibility in processing. Not only can grain be transported great distances, but long term storage possibilities allow for year round processing of recombinant protein. Downstream processing can be matched with consumer demand thereby reducing inventory costs, especially for perishable proteins.

The typical bioprocess for transgenic corn is depicted in Figure 3. Transgenic corn is harvested by using dedicated harvesting equipment (combines) and transported to the storage facility. If necessary, transgenic grain can be cleaned and dried under specified conditions before storing. The recombinant protein in the stored transgenic grain is stable for at least a year and can be inventoried to allow year-round protein production. As needed, the transgenic grain is transported from storage to the bioprocessing facility to be degermed and /or ground to a prescribed particle size before protein extraction.

Ground grain or germ, referred to as meal or flour depending on particle size, is then extracted with buffer or water. Extracted corn meal is separated by centrifugation or filtration to produce a clarified crude protein extract. The extraction of corn meal/flour is very flexible and allows the optimization of the processing conditions (water vs. buffer, temperature, pH, mixing time, etc.) to achieve the maximal recombinant protein yield. For example, a soft water adjusted to a required pH with acid or base could be used in the extraction, resulting in a significant reduction of the overall processing cost.

Depending on the end use and final product specifications, the protein extract is further purified by using one or more purification steps such as precipitation, membrane filtration, and chromatography. For example, highly purified proteins for pharmaceutical applications will require the use of at least two purification steps including chromatography. The production of industrial proteins typically requires grinding and protein extraction, whereas products to be directly delivered as flour, such as human or animal oral vaccines, will need only dry grinding and formulation. Products that can utilize minimally processed plant fractions will enjoy a significant cost advantage. It is important to note that regardless of the end use, the same front-end infrastructure and grain processing technology will be used (grinding, degermination, wet milling, etc.). Furthermore, corn byproducts such as starch and oil may be salable as chemical and fermentation feedstock or even animal feed if the expressed protein is proven to be safe.

PRODUCT COST CONSIDERATION

The decision-making process for choosing the best production system is a complex one, requiring a case-by-case analysis. Factors to be considered when selecting the most suitable transgenic plant systems include production cost and volume, end product application (pharmaceutical vs. industrial protein), product specification, speed to market and regulatory constraints. From a bioprocessing viewpoint the following criteria could be applied in case analyses (Cramer *et al.*, 2000; Pen, 1996; Krebbers *et al.*, 1992; Kusnadi *et al.*, 1997):

- An option for directing protein accumulation to a harvestable tissue/organ to assure protein quality, quantity, and easy recovery
- Crop/protein yield and cost
- Protein accumulation and stability in harvested tissue
- Protein concentration and stability in the extract
- Extract properties and complexity
- Applicability of standard purification methods
- Disposal cost vs. byproduct revenues

The applicable criteria and their weight in the evaluation process are product dependent and may not apply across all protein products and plant systems. For instance, the considerations for large-volume, low-cost protein products such as industrial enzymes that require no purification will be different than those applied to biopharmaceuticals.

To demonstrate the influence of different criteria in the evaluation of these two types of products, the manufacturing cost breakdown of a simulated corn bioprocess for producing an industrial protein and a pharmaceutical protein is shown in Figure 4. The design and simulation of the processes outlined in Figure 3 were performed using Super Pro-Designer 4.5 software (Intelligen, Inc., Scotch Plains, NJ). The cost breakdown for a highly purified pharmaceutical protein from transgenic corn shows that the transgenic raw material constitutes only 5-10 % of the total manufacturing cost, whereas extraction and purification make up as much as 90% (Figure 4). By contrast, the cost of ground transgenic corn for a typical industrial protein, which is extracted and then concentrated by ultrafiltration, amounts to 35 - 40% of the total manufacturing cost. Therefore, for industrial products where large-volume, low-cost product is the primary concern, crop yield, protein accumulation in harvestable storage tissues, infrastructure for handling and processing of crop species and capturing byproduct revenues are critical for favorable economics. For biopharmaceutical proteins, on the other hand, subcellular localization to ensure correct processing will be of primary importance. However, considerations such as recombinant protein concentration and targeting expression to crops or tissues containing the fewest interfering impurities that may adversely affect protein purification will also figure prominently in the decision making process.

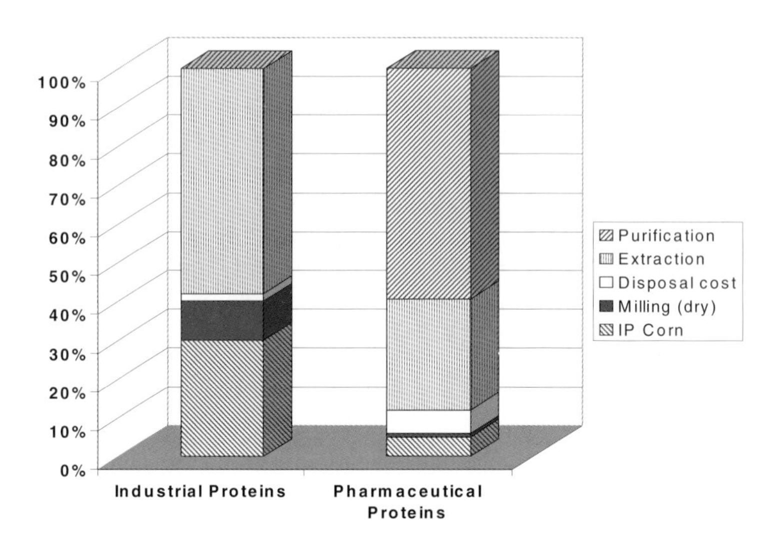

Figure 4. *Distribution of Manufacturing Costs*

Clearly, transgenic material costs for some products will be more important than for others. Several potential strategies may be applied to reduce raw material cost for lower value plant-based protein products. These include:

- Increasing the concentration of the target protein in the raw material by harvesting part of the plant or by fractionating the harvested tissue which contains the greatest amount target protein
- Co-expressing multiple proteins in a single plant variety
- Stacking traits such as combining a recombinant protein with an oil or carbohydrate quality trait
- Maximizing yields and agronomic performance of the transgenic variety or minimizing special management practices required by growers
- Generating and utilizing crop byproducts. The absence of a feed and food tolerance for the expressed product may limit byproduct utilization. For corn, fractionation of the grain and utilization of byproducts for feed or fermentation feedstock may be a viable strategy, if feed tolerance is established

SUMMARY

As we have seen, many factors come into play when deciding on a plant system for the production of recombinant proteins. To be competitive with other production systems, costs must be kept to a minimum for all but the most expensive pharmaceutical proteins. Crop production considerations such as plant productivity, isolation, IP systems and regulatory compliance must be carefully balanced against processing concerns such as harvest, storage, purification and byproduct disposal. Furthermore, each step of the process must be carefully monitored with built-in safeguards to ensure product quality and containment. Ultimately, the end product and its application will determine which approach is the most feasible and cost effective. While relatively few plant systems are currently ready for industrial scale protein production, research into new systems is ongoing. The current discussion, while limited to tobacco, alfalfa and corn, will hopefully prove useful in evaluating emerging plant technologies as well.

REFERENCES

Austin S, Bingham ET, Koegel RG, Mathews DE, Shahan MN, Straub RJ and Burgess RR 1994. An overview of a feasibility study for the production of industrial enzymes in transgenic alfalfa. *Ann. NY Acad. Sci.* 721, 235-244.

Cramer CL, Boothe JG and Oishi KK. 2000. Transgenic plants for therapeutic proteins: Linking upstream and downstream strategies. In *Plant Biotechnology* (ed. J. Hammond, P. McGarvey and V. Yusibov), pp. 95-118. Springer, New York.

Doran P 2000. Foreign protein production in plant tissue culture. *Current Opinion in Biotechnology* 11, 199-204.

Fiedler U, Phillips J, Artsaenko O and Conrad, U. 1997. Optimization of scFv antibody production in transgenic plants. *Immunotechnology* 3, 205-16.

Fischer R, Drossard J, Commandeurr U, Schillberg S and Emans N. 1999. Towards molecular farming in the future: moving from diagnostic protein and antibody production in microbes to plants. *Biotechnology & Applied Biochemistry* 30, 101-108.

Garger SJ, Holtz RB, McCulloch MJ and Turpen TH. 2000. Process for isolating and purifying viruses, soluble proteins and peptides from plant sources. 6,033,895. Biosource Technologies, Inc., USA.

Gegenheimer P. 1990. Preparation of extracts from plants. *Methods in Enzymology* 182, 174-193.

Hammond, J. 2000. Shed and Spread of Transgenes. In *Proceedings: Plant Derived Biologics Seminar*, pp. 39-46, Ames, IA.

Hanna M and Greenless W. 2000. Planter clean-out tips when changing seed varieties. Iowa State University website.

Henderson L. 2000. Regulatory Considerations for Licensure of Plant-Based Vaccines. In *Proceedings: Plant Derived Biologics Seminar*, pp. 76-83, Ames, IA.

Krebbers E, Bosch D and Vandekerckhove, J. 1992. Prospects and Progress in the Production of Foreign Proteins and Peptides in Plants. In *Plant Protein Engineering* (ed. P. R. Shewry and S. Gutteridge), pp. 315-325. Cambridge University Press, New York.

Kusnadi AR, Hood E, Witcher D, Howard J and Nikolov Z. 1998. Production and Purification of Two Recombinant Proteins from Transgenic Corn. *Biotechnol. Prog.* 14, 149-155.

Kusnadi AR, Nikolov ZL and Howard JA. 1997. Production of Recombinant Proteins in Transgenic Plants: Practical Considerations. *Biotechnology and Bioengineering* 56, 473-484.

Loomis WD. 1974. Overcoming problems of phenolics and quinones in the isolation of plant enzymes and organelles. *Methods in Enzymology* 31 (Pt A), 528-544.

Pen J. 1996. Comparison of host systems for the production of recombinant proteins. In: Owen MRL, Pen J (eds) Transgenic plants: a production sytem for industrial and pharmaceutical proteins. Wiley, Chichester, pp 149-168.

Sumere CF v, Albrecht J, Dedoneder A, Pooter H d and Pe I. 1975. Plant Proteins and Phenolics. In *The Chemistry and Biochemistry of Plant Proteins* (ed. J. B. Harborne and v. Sumere), pp. 211-264. Academic Press, London.

Theisen M. 1999. Production of recombinant blood factors in transgenic plants. In *Chemicals via Higher Plant Bioengineering* (ed. F. Shahidi, P. Kolodziejczyk, JR. Whitaker, AL. Manguia and G. Fuller), pp. 211-220. Kluwer Academic/Plenum Publishers, New York.

Turpen TH. 1999. Tobacco mosaic virus and the virescence of biotechnology. *Philosophical Transactions of the Royal Society of London - Series B: Biological Sciences* 354, 665-673.

USDA/APHIS. 2001. www.aphis.usda.gov/biotech/7crf340.html#340.3.

White J. 2000. APHIS' Experience in the Field of Testing of Plant-derived Biologics. In *Proceedings: Plant Derived Biologics Seminar*, pp. 69-75, Ames, IA.

Yildirim S, Fuentes RG, Evangelista R and Nikolov ZL. 2001. Fractionation of transgenic corn for optimal recovery of recombinant enzymes. *J. Am. Oil Chem. Soc.* submitted.

REGULATORY CONSIDERATIONS IN A CHANGING ENVIRONMENT

Don Emlay

EMLAY and Associates
416 Carmel Avenue
Pacific Grove, CA 93950
Demlay@earthlink.net

INTRODUCTION

There is a transition underway in transgenic plants. The major transgenic plants commercialized to date have expressed input traits in commodity crops intended as food and feed, grown over large acreages and approved for food and feed uses prior to commercialization. The new transgenic plants being developed and commercialized will produce high-value output products that are not intended for use as food or feed. The types of expression products range from industrial enzymes to human drugs and biologics, with product forms from edible vaccines to highly purified pharmaceuticals. These products and the crops in which they are produced will be different from crops that are commodities or are engineered to express new food products. Consequently, regulatory considerations will be fundamentally different. As a result, these new crops will need to be grown, handled, managed and processed differently from current transgenic crops. Additionally, public perception places the industry under increased scrutiny. As awareness increases that human and animal pharmaceutical products are now being produced in crops that have to date only been used as human food and animal feed, regulatory agencies and the industry are being asked to demonstrate that measures are being taken to prevent an occurrence such as the inadvertent introduction of Aventis Crop Science's StarLink brand corn into the human food chain. It is evident from the StarLink incident that even a single similar incident with a pharmaceutical plant product would be devastating to the company involved and would have a significant negative impact on other companies developing similar products as well as transgenic plants in general. While it is believed that every company developing new transgenic plant products understands the importance of having procedures to ensure that the product is "contained" during all stages of production and

E.E. Hood and J.A. Howard (eds.), Plants as Factories for Protein Production, 175–180.

processing, the need for total containment procedures during development and commercial production introduces a complex element that impacts the regulatory process and every aspect of commercialization. While containment procedures are intended to prevent the presence of the expression product in food and feed, many of the same procedures are applicable to maintaining the integrity of the growing crop to ensure that only the intended expression product is present. Although intended to accomplish two different goals, the procedures to ensure containment and product integrity are linked and are likely to become part of the documentation supporting product approvals. The need for procedures for containment and product integrity is indicative of the complex issues surrounding the use of transgenic plants to produce certain types of products. For some products (e.g. industrial enzymes or other non-pharmaceutical products) containment may be the dominant element, with product integrity being less important. On the other hand, containment and integrity are of equal importance in the production of human pharmaceuticals in plants.

The regulatory considerations, data requirements and regulatory processes for transgenic plants and pharmaceuticals, each taken separately, are well defined. The USDA and FDA have provided extensive guidance in these areas. It is the combination of the two that creates the area that is currently being examined by regulatory agencies and those companies developing these new products. Even with this situation there is guidance that can steer a company in the right direction. This includes the results of USDA sponsored meetings[1], industry sponsored workgroups[2], and numerous Points to Consider documents available from the FDA. The Draft White Paper Prepared by the Human Therapeutics in Transgenic Plants Industry Group provides guidance to understanding the specific data that are likely to be required to support product safety. While many of the areas in which data are required for plant-based pharmaceuticals continue to be examined and refined, the most interesting and perhaps most contentious area of producing human pharmaceuticals in plants is associated with moving a portion of the manufacturing process to an open field where variables are a constant factor. This is the major area on which this chapter will focus.

---

[1] Regulatory Considerations for Licensure of Plant-Based Vaccines, Center for Veterinary Biologics, Ames, Iowa

[2] The Manufacture and Testing of Medicinal Biological Products for Human Use as Derived from Transgenic Plants, Draft White Paper, Human Therapeutics in Transgenic Plants Industry Group

The Manufacturing Process and Transgenic Plants

Defining the manufacturing process is perhaps the most important and difficult area of regulatory requirements for pharmaceutical-expressing plants. This starts with the differences between producing a pharmaceutical product in a "closed system" as compared to having a major part of the production process occurring in an uncontrolled environment or "open system" and includes the application of cGMP (current Good Manufacturing Practices) procedures. A well-defined and consistent manufacturing process for pharmaceutical products is a major component of a regulatory approval. It is generally recognized that even subtle changes in the process can result in minor but undesirable changes in the finished product. The fact that pharmaceutical products are now being manufactured in transgenic plants in an environment where controls are limited, introduces a unique element into the manufacturing process that must now be considered by regulatory agencies and the companies developing these products. Essentially, the processes and controls represented by a cGMP facility are missing during much of the manufacturing process with transgenic plant products. The definition of an acceptable manufacturing process is somewhere between the need to have complete control over the manufacturing process and acceptance of the conditions that exist with growing plants.

One approach is to consider the manufacturing process as four major components, (1) data and procedures to support the consistent expression of the intended product, (2) field production procedures that are uniformly applied and recorded to ensure containment and product integrity, (3) pre-processing of the harvested plant material in a controlled environment and (4) production of the finished product in a cGMP facility. The application of this approach requires the FDA to accept the variables associated with growing plants and the developing company to implement procedures that have never before been applied to a growing crop.

Data and procedures to support the consistent expression of the intended product starts with data that are presently generated and required by the FDA to demonstrate the stable integration of the intended gene into the plant genome and expression of the intended product from generation to generation. These basic requirements are presented and discussed in the Human Therapeutics in Transgenic Plants Industry Group Draft White Paper. The need to be absolutely certain that the intended expression product is unchanged from generation to generation and between production locations may require additional testing not defined in current guidance. Companies may want to consider introducing procedures to monitor the expression product at different stages of production to ensure it is present as the intended product prior to sending a partially processed plant material to a cGMP facility.

Field production procedures that are uniformly applied and recorded to ensure containment and product integrity must be developed and implemented. Control procedures involved with production are briefly discussed in subsequent paragraphs and in greater detail in Chapter X. While variables related to the environment are not controllable in an open field, many of the procedures associated with growing a crop can be controlled. These controllable procedures are primarily related to isolation of the crop, agronomic practices and crop handling procedures.

Procedures to ensure containment and product integrity may be considered in two phases, (1) during the development stage prior to any regulatory approvals and (2) during commercial production after regulatory approvals. During the development stage, it is imperative that the transgene and expression product are confined to the specific transgenic crop. Containment requirements during commercial production will be directly dependent upon the nature of the expression product and intended uses and approvals obtained for the expression product. The procedures to maintain product integrity are similar to the Identity Preservation procedures used today with specialty crops but must be more precise. These procedures are essentially expanded containment procedures and are necessary to ensure that expression products from other transgenic crops are not mixed with the intended crop through pollen transfer or through inadvertent mixing with the harvested crop. The procedures to ensure product integrity are especially important in nursery situations where multiple transgenic lines may be present. Plant lines that are selected from a nursery for further development should be screened to ensure only the intended genetic sequences are present. Isolation procedures must be applied to the crop during all production phases. This means that the procedures to ensure the transgenic crop is not mixed with commodity food or feed will also serve to prevent the mixing of commodity food and feed crops with the transgenic crop. These procedures will most likely become part of the manufacturing portion of the documentation submitted in support of a regulatory approval. Procedures to ensure containment and product integrity represent one means of reducing the variables in a specific growing site.

Reducing variables due to agronomic practices is a function of applying the same procedures to the same product from production site to production site and from season to season to the maximum degree possible. It will be important for regulatory agencies to recognize and accept that chemical usage will vary between growing regions due to different pests and that the application of water and fertilizers are affected in the same way. Record keeping and absolute compliance with labels will be a critical element of this aspect of production.

As with the procedures to ensure isolation, crop-handling procedures also serve to contain the crop and maintain the integrity of the intended expression product. These procedures must be applied to every step and phase of developing and producing a pharmaceutical-expressing plant. This includes records to confirm and trace the history of every experimental or production field and the implementation of procedures to definitively identify the product during all stages of production. In addition, planting, harvesting, transportation, storage and processing equipment that is dedicated to the crop or can be thoroughly cleaned before, after and between uses is critical. Perhaps the most significant element of this aspect of production is a complete understanding and acceptance of these procedures by all those involved with the growing and processing of these products.

Pre-processing of the harvested plant material in a controlled environment is a critical component of the manufacturing process. This is a production stage that may be compared to the introduction of raw materials into the cGMP facility. This step involves a consideration of the form of the product that will subsequently enter a cGMP facility for production of the finished product. For example, if corn kernels will be initially processed to produce a concentrated mixture of corn proteins and the intended product, this step could take place in a sterile environment similar to that used in food production. This step with the proper procedures and controls may serve as a bridge between the field production component and the cGMP procedures.

Production of the finished product in a cGMP facility is a component of a major debate currently taking place. This debate surrounds identifying the stage in the production of pharmaceutical-expressing plants at which cGMP procedures should be applied. The entire manufacturing process for pharmaceutical products produced by traditional methods is conducted under cGMP procedures in a closed system. The successful commercialization of plant-based pharmaceutical products depends, to a great part, on where in the production process cGMP procedures start. Perhaps the current production of pharmaceutical products from naturally occurring plants (e.g. bark from *Taxus brevifolia* processed under cGMP conditions to produce Taxol) can be used as precedents for policies relating to this area. The most relevant factor is that the highly purified pharmaceutical produced in the cGMP facility is identical to the approved product and is the same from batch to batch. There is significant scientific evidence that a gene inserted into a plant will consistently express the intended product. The expression level will vary, as this is influenced by many factors, but methods are available to ensure the plant consistently produces the intended expression product. Quality assurance procedures applied to the finished product can detect changes to the product in the same manner as they are now used with traditionally produced pharmaceutical products.

CONCLUSION

The majority of the opinions expressed in this chapter are predicated on the assumption that the procedures associated with the growing of a pharmaceutical-expressing plant will become part of the documented manufacturing process required for a product's approval. If this assumption is correct (and to keep the cGMP element within the facility where the highly purified product is made), the procedures applied to the pre-cGMP production processes must be precise and fail-safe. These procedures need to be developed and implemented when the first transformant is made. Considering how critical these procedures are to the manufacturing of pharmaceutical products from plants, it is not unreasonable to anticipate formal guidance in this area from the FDA to establish industry standards.

Over the past few years there have been conferences sponsored by the FDA and USDA to discuss new transgenic plant products and potential changes in data requirements and regulatory processes. An industry sponsored working group (refer to footnote 3) is preparing a document to serve as guidance to both industry and agencies for human biologics expressed in plants. It is important that companies developing new plant products attend these conferences and participate in this and other working groups. This involvement can help shape the decisions and direction of the agencies and provide valuable insight into how the agencies are thinking about these products. It is likely that a company working on a new product today will require a commercialization plan that includes regulatory requirements that will have to be identified well in advance of published guidance from agencies and will therefore require an understanding of what the differences are and how they will affect agency thinking and the approval process.

A Warehouse of Ideas; Developing and Using Intellectual Property

Patricia A. Sweeney, Esq.
1835 Pleasant Street
West Des Moines, IA 50265-2334

"Though this be madness, yet there is method in't."
William Shakespeare, Hamlet

INTRODUCTION

If leading lawyers can be metaphorically described as trying to herd cats, then attempting to understand and make intelligent use of intellectual property laws for a business person is not unlike trying to understand the entire topography of a river by taking out a cup of water and studying it. It never stops changing, and each corner, each situation carries with it its own peculiarities. It is particularly true for those in the biotechnology industry. When, for example, one needs to understand the ability to work on a particular project free of the threat of infringing other's patents, searches are conducted, patents and applications evaluated, and a picture is painted of what the intellectual property landscape looks like. Yet, every Tuesday new patents issue from the U.S. Patent Office, and the entire picture could change. Further, just how the courts or patent office interprets the requirements and rules for obtaining a patent are in constant evolution, and what was true several years ago may now have turned completely around.

Nor are the stakes small. For some companies, their intellectual property is their asset. There was a time when a business person could visit a warehouse and see and count the key assets. Now, there may be no buildings, and only a few principals. Yet companies can sell for millions based on these intangible assets alone. Further, patent infringement issues can sting. Consider verdicts like the $65 million damages awarded Rhone-Poulenc for claimed infringement by DeKalb of its patent to technology related to herbicide resistant corn, or the $174.9 million awarded to Mycogen based on its claim that Monsanto did not fulfill terms of licensing of crop enhancement technology.

Thus, any important perspective for the business person to keep in mind regarding intellectual property in the plant protein production field is this: it's important, it must be a key part of the original business plan and focus, and to understand that what one has at the end of a freedom to operate or state of the art opinion is a snapshot that may change next week or even tomorrow.

Therefore, the following is not intended to provide discussion of key intellectual property holdings in the area of plant protein production. Such a review would be outdated before it went to press. Instead, this chapter aims to provide an overview and potential game plan for creating your own portfolio of value in intellectual property while attempting to avoid the pitfalls of wandering into infringement of another's intellectual property.

There are several views, which must be taken to obtain a broad picture of intellectual property and how it can affect your company. What is in your own patent portfolio? Do you have "freedom to operate"? Comparing it to building your own home, it is not unlike asking what kind of house you have; is there a good foundation? Does it have a variety of useful features? What's the best way to build it? Also, do you properly own the deed? Does anyone else have ownership rights to the house? What are the access roads like, can you get to your house without problems?

CREATING YOUR OWN INTELLECTUAL PROPERTY PORTFOLIO

The earlier you begin building your house, to carry out the metaphor, the better and the earlier a company focuses on intellectual property, the better positioned it likely will be. This often isn't easy since money can be tight at these stages, but even if the company is still being run out of your basement, it's not too early to consider your intellectual property. Train scientists early to consider what is patentable. Provide lab notebooks and teach the scientists how to properly keep them. Have a person, if at all possible, who is devoted to looking for and shepherding intellectual property issues. It can pay off in the long run. The following gives an overview of some of the key issues in developing your intellectual property.

What are the options?

Intellectual Property is a term broadly including patents, trade secrets, plant variety protection (PVP), trademarks and copyrights. It is the first three that will be the focus of this review. Each one has its own advantages and disadvantages. The table below provides a summary that compares features of each. Note that publication of the invention is included; it is not "protection" per se, but, as discussed below, is a defensive alternative in attempts to keep the area open and available for us.

Table 1. *Summary of Intellectual Property Protection Options*

	Patents	Trade Secrets	Publication	PVP (if plants)
Difficulty to obtain	Overall difficult Requires: - Novelty -Non-obvious Patentable subject matter	Overall moderate Requires: - Derives value from being secret - "Reasonable" precautions	*	Overall easy Requires: - Distinct - Uniform - Stable
Time to obtain	2-5 years	Immediate	*	About 1 year
Cost	Costly	Varies	*	Moderate
Term	17/20 years	Until secret is lost	*	20 years
Remedy	Control making, using, and selling	Stops misappropriations	*	Stops reproduction of the variety
Controlling Law	Uniform	State law (50 options)	*	Country law
Best Benefit	"Cadillac" protection	Easy/indefinite term	Is art to those who invent after publication	Fast, improved protection; low cost
Problems	Exposure after 18 months unless filed in the U.S. only	Reverse engineering allowed; independent invention allowed	Not really "protection", is a defensive action	Breeders exemption; indefinite definitions

Patents

Patents are considered the "cadillac" of intellectual property protection; they can provide broad coverage of an invention and enforcement has teeth. The idea behind patent protection is that, if the patent meets the standard of patentability, the owner is allowed at most 20 years, and possibly shorter protection[3] in the form of the ability to decide who can make, use and sell the invention. What is it that the owner can stop another from making, using or selling? The invention *as it is claimed.* Claims are the numbered description of the invention at the end of the patent. Claims to a patent are like a metes and

[3] In general, a patent filed and issued before June 8, 1995 has a term of 17 years from issue; if pending on June 5, 1995, it has the longer of 17 years from issue or 20 years from filing; if filed after this date it has a term of 20 years from filing. The term can also be shortened or changed by virtue of a number of different circumstances.

bounds description is to a piece of property. Therefore, claim language in a patent is often convoluted and downright strange. This is because the patent attorney is attempting to write the description of the invention in the claims broadly enough that no one can try to escape the scope of the claim, but narrow enough so that it will not be invalid because it is too broad. Thus, if another's product or process falls within the scope of the claims, they can be sued for infringement of the patent by the owner. Compensatory damages are possible, which can take a number of forms, including lost profits or payment of a reasonable royalty. Further, if the infringement is willful the damages award can be trebled. Thus, it can be a potent tool.

A disadvantage of patents is that in order to obtain this tool, the standards of patentability must be met, and this is not always easy. The U.S. patent laws require the invention be measured in light of prior art; that is, on the date of the invention, one looks at all that was publicly known before. Without going into detail, prior art includes public disclosures, such as talks at conferences, publications, published patent applications, theses, and the like. It also includes offers for sale of the invention that occur more than one year before the filing of the application. (Note that European law differs, as discussed later.) An invention must be 1) novel or new in light of the prior art, 2) non-obvious in light of the prior art, 3) have utility, and 4) be patentable subject matter. To be new, the invention must not be found described within any one prior art reference. To be nonobvious, the invention must not be an obvious combination of any of the prior art disclosures. Utility can be an issue where the patent office doesn't believe the invention will work (such as a cure for cancer) and subject matter is rarely an issue these days. It may be difficult to convince the Patent Office that an invention deserves patentability. It can take two to five years, or more, to obtain a patent. Nor is it an inexpensive process, when adding the Patent Office fees and attorney fees.

Further, a patent is granted in exchange for full and complete disclosure of the invention. The purpose behind the patent laws is to encourage disclosure of ideas that otherwise might be kept secret. The right to decide who makes, uses or sells the invention for a period of time is the reward for the disclosure; hence a key issue in obtaining a patent is that the disclosure teaches others how to practice the invention. Trade secret protection, on the other hand, is provided only when the invention is kept secret and obtains value as a result of that secrecy, and where reasonable precautions are used to protect that secrecy. Thus, one cannot have both types of protection on the same aspect of an idea. At the same time, it is possible to have trade secret protection while a patent application is pending, and in this respect there can be some overlap. Up until eighteen months from the filing of an application that will be also filed outside the U.S., the patent application is secret. If the application is to be filed outside of the U.S., the application will publish at the end of the year-and-a-half. At that time, trade secret protection will be lost. Thus, one can use trade secret protection up until that period (provided the information is not otherwise disclosed) and decide before publication if it has more value as a trade secret or as a patent. If no protection outside the U.S. is sought, the patent application will remain secret until it becomes a patent.

Trade Secrets

Trade secret protection can be much less expensive; it depends on the steps that will be needed to keep the information secret. The precautions to keep the information from getting out must be "reasonable". Also, unlike patent protection, trade secret protection lasts as long as it can be kept secret. Further, much more can be protected than with patents. In general, the trade secret must derive value from being secret, and reasonable precautions must be taken to keep it secret. A wide variety of information is potentially protectable: business record systems, laboratory methods, client lists, etc. Thus, it is usually easier to obtain, costs less, and lasts longer. One also need not wait for any approval before having trade secret protection.

Given these advantages, what are the down sides? One can stop another from misappropriating the idea, but cannot stop someone from obtaining it legitimately. If it is possible to reverse engineer the invention; that is to tell what the idea is by studying an end product, this is permitted. Further, if another independently comes up with the idea without misappropriating it, this too is permitted. Finally, if the idea somehow gets out through an inadvertent disclosure, it's gone and the cat can't be put back in the bag. Thus, it's risky. One other aspect of trade secret law to note is that it is not a federal law, as with patents. Instead, each state has its own version of trade secret protection, and a fair amount of uncertainty accompanies this.

Plant Variety Protection

When involved in production of proteins in plants, there is yet one more type of protection available. The Plant Variety Protection Act provides protection for a particular *variety* of a plant. In other words, if corn is the crop used for production of the protein, and a particular inbred or hybrid produces exceptional results, that variety can be protected through the PVPA. Once a PVP Certificate is granted, it is public, so trade secret protection is not possible; however, in the United States, one can have both PVP protection and patent protection.[4] Also, the European Patent Office determined that it was possible to obtain patent protection for plants. [5] They also allow PVP-type protection. In nearly every other country, however, it is not possible to obtain a patent to a plant. Patents can be obtained for a new protein, new method of obtaining a protein, a new composition, isolated nucleotide sequences, vectors or the like. However, if one wants to protect plants in countries other than the U.S. and Europe, in general, the only protection available will be Plant Variety type protection.

There is an important exemption to infringement, however. If the variety is being used to develop a new variety, this is permissible. It is reproduction of the variety that is prohibited. Previously, an exemption also allowed farmers to save seed produced from the protected variety seed, but this has now been changed. The farmer can save seed for another planting, but not to sell to another.

Also, the PVPA has added that one may infringe a protected variety if their variety is essentially derived from the protected variety. This is a variety that is derived from the protected variety and 1) retains the expression of the essential characteristics that result from the genotype or combination of genotypes of the initial variety 2) is clearly distinguishable from the initial variety except for differences that result from the act of derivation and 3) conforms to the initial variety in the expression of the essential characteristics that result from the genotype or combination of genotypes of the variety.[6] Just what is "too close" for the purposes of determining if another variety is essentially derived from another variety? It's uncertain because there have been no test cases and little guidance from the drafters of the amendments to the statute. Thus, this area of uncertainty in the PVP law remains.

Defensive Publication

Finally, what if patent protection is not available, trade secret protection is not practical? In that case, it might be wise to consider publication as a defensive action. When the invention is published, it becomes prior art against those who might want to later patent the idea, and also stops

[4] Note that in the case of *Pioneer Hi-Bred v. J.E. M. Ag Supply, Inc.*, 2000 U.S. App. Lexis 682 (Fed. Cir. 2000) the Federal Circuit Court decided that plants could be provided both patent and PVP protection. At this time the parties had applied for certification to the Supreme Court.

[5] See Decision of the Enlarged Board of Appeal of the European Patent Office G1/98 Dec. 20, 1999 and Rules 23b-e EPC, OJ EPO 1999, p.576.

[6] 7 U.S.C. §2543.

others from using it as a trade secret. Stopping another from obtaining patent protection will only be effective if the other party's invention is created after the publication. However, it can be a wise move to at least keep the area open for your own work.

Which to Choose?

Some practical considerations need to be kept in mind when determining which option is optimal.

Patents: Whether and When to File

In determining whether patent protection fits the invention, consider the following:

- How likely can infringement be detected?
  - ➢ If a lab process is involved, for example, might you obtain a patent that could be infringed, but you would never know? Or, if someone were to use the idea, would they necessarily have to mention this in any publications of their work? In the latter, monitoring publications might be worthwhile. However, if there's little chance of knowing when someone is infringing, patenting might give you a tool you could never use.
  - ➢ At the same time, even if detecting infringement might not be high, the patent could have a deterrent affect upon others.
- What scope of protection can you get?
  - ➢ Are the claims likely to be broad, intermediate or narrow? A pioneering invention can provide considerable value to a company, but remember that narrow claims may still be quite valuable, especially if the claims describe the most efficient, best means of obtaining product or results. This is discussed further below.
- Value vs. Term of protection
  - ➢ The time from filing to patent issuance can vary greatly. On the shorter end of the spectrum, it is possible to obtain protection within a year, but the broader the claims, the less likely that will occur. Instead, two to five years is not uncommon. An important consideration is whether the value of the invention will have expired before the patent issues. Where there is fast turnover, patent protection may not be practical.

- Compare expense of obtaining patent protection to end use.
  - ➢ How you might use your patent will impact whether it is worth the expense. Your patent and marketing advisors can help you predict whether the patent could be used for any of the following: 1) keeping an area exclusive to yourself; 2) providing a head start; 3) making it more difficult for others to practice in your area; 4) licensing out to others; 5) trading for tools you need; 6) making sure you can continue to use your own idea. Also, how close is it to the actual product? Does it cover a component, process or equipment used to make the product? Is it strictly a research tool? Does it cover the product itself?
- Do you have enough substance to warrant a patent?
  - ➢ It can be a risky proposition to file a patent application when you don't have enough data or information to support your patent. Why? Once the patent application publishes, it will be prior art against your later work as well as against others. Thus, you could prevent later patenting with your own earlier weaker application. For example, suppose a patent application publishes that teaches that a particular gene group can be expressed in plants, when previously no one thought it was possible. However, you have little data and do not adequately teach others how to make the invention work. Later, you find out you were right. However, you can no longer obtain broad claims to the gene group expressed in plants; instead you must narrow your claims to that particular aspect not suggested by the prior publication.
  - ➢ Thus questions to ask include 1) do you teach one skilled how to make and use the invention; 2) do you now enough to repeat the invention and get the results you want; 3) can you provide examples; and 4) do you have enough to support the breadth of claims you want?

Provisional Applications. A tool provided by the Patent Office offers help in this area: the provisional application. This type of filing lets you place a "bookmark" in time for your invention. Such filings are very simple, and often no more than a publication draft. They do not advance through the Patent Office, and instead are inactive for one year following the filing. At that time, the patent application must be converted to a conventional filing, and international filings that are desired must also occur at the end of the year. What that does, however, is buying a year of time to further develop the invention. More than one provisional can be filed during that time, and the final conventional application can relate back to all of them. Therefore, if you may have enough data within a year, it may be worthwhile to file a provisional application. If you are not sure whether you will have the data and/or information within a year, the provisional can be dropped before conversion.

Keep in mind what kinds of data you are waiting for before deciding whether a provisional is appropriate. Do you have enough data to convince the Patent Office you knew what the invention was and had it in hand? If not, you may face a 35 U.S.C. section 112 second paragraph rejection for lack of written description, in which the Patent Office indicates, in essence, it doesn't believe you really had the invention and knew it well enough when you filed. Do you have enough information to teach the invention to another? If not, you could receive a section 112 first paragraph rejection that your application is not enabling. In this regard, it is important to note that it is possible to add what are called prophetic examples to the application. These are examples of experiments you plan to conduct, but haven't yet. Such examples can be useful *if* they work out later as you plan. Also, there are no examples, prophetic or otherwise, that will save an application that doesn't teach the invention.

Importance of Adequate Disclosure. This requirement of showing that you have the invention at the time of filing has become increasingly important with decisions from the Federal Circuit. For example, in *University of California v. Eli Lilly & Co.*, 119 F3d 1559 (Fed. Cir. 1997) the Court found a patent invalid where it described rat insulin cDNA and amino acid sequences of human insulin A and B chains. However, the claim in question was to a microorganism modified to contain a nucleotide sequence having the reverse transcript of an mRNA of a human, encoding insulin. The court said no information was provided on the structure or physical characteristics of human nucleotide sequences as described. This claim and the application did not meet the written description requirements. This was also the case in *Amgen v. Chugai*, 927 F2d 1288 (Fed. Cir. 1991) where the applicant attempted to claim all sequences encoding a protein sufficiently similar to native human erythropoietin (EPO) to be biologically active. The court found a lack of written description since more than 3600 different EPO analogs were possible with just one amino acid substitution.

Regarding enablement, the court in *Genetech, Inc. v. Novo Nordisk* 108 F3d 1361 (Fed. Cir. 1997) went back on earlier concepts that a patent application need not teach what is already known, and doesn't need to be a treatise on a particular subject. Instead, in vacating a preliminary injunction against Novo, they found that the application in describing a method of producing human growth hormone using cleavable fusion expression did not meet the enablement standard. Genetech argued that a person skilled in this area would know how to use cleavable fusion expression to make the hormone, and that they did not need to go into detail on this or specific reaction conditions when describing a method of transforming bacteria with DNA encoding human growth hormone and cleaving the resultant conjugate protein. The court responded that the more unpredictable the area of science is, the more that must be described, and here it was not enough. Just how much is enough will be determined in the future by trial and error; but it does cast concern on just when a patent applicant has enough to file an application.

In Summary. Thus, one is faced with competing interests; you want to file as early as possible, but don't want to file a patent application so weak that if it published it would damage you later. You don't have to know how your invention works, although theory is useful in arguing a case with the Patent Office. You do have to know what to do to make it work. You technically don't even have to provide an example, but with the recent court decisions, it would be unwise to file without at least a prophetic example.

If you believe you have enough to teach others how to recreate the invention, but expect the Patent Office will ask for more supporting data later, a provisional application can be most useful when you expect the data within a year. If the data may be forthcoming in four to six months, you may want to file a conventional application and wait for the first action from the Patent Office. How long this takes can vary even from year to year depending on the appropriations bills passed in the legislature for the Patent Office, but it is often six months at least for the first office action. In this time, you can marshall your supporting data. However, again make sure that if you file a conventional application you have taught enough that others can practice the invention; the data you will be waiting for merely confirms what you have already taught.

Trade Secrets as an Alternative

As mentioned above, it may be possible to use trade secret protection for an invention even while applying for a patent application, for the first eighteen months. This will only work, of course, if you prevent any publications or other public disclosures until patent publication. However, except for this type of situation, one ordinarily must choose between patent and trade secret protection. Trade secret protection may seem attractive where the invention does not meet patentability standards, is not worth the expense of a patent, or where its life span of value contribution far exceeds the length of patent protection. There are several practical considerations when determining if trade secret protection is the right alternative:

- Is it practical to keep it secret?
    - ➢ Some ideas or information are just too difficult to keep secret. It may be too simple and easy to reveal inadvertently, may be exposed to too many people, and may need to be revealed to support a patent or publication, or require too much expense to keep secret. Some companies go to great lengths to keep their trade secrets; one company for example inspects every briefcase going out of the building. Here again, the value of the secret versus the costs must be compared.
- Is reverse engineering possible?
    - ➢ If it is possible for someone to determine the idea, information, or invention by taking a publicly available document or product, such reverse engineering is allowed. In that case, a competitor will be free to use the invention. Further, they could also reveal it publicly, defeating your trade secret protection.
- Is independent invention likely?
    - ➢ Could someone come up with the same idea? If so, they are allowed to use the idea and you could not stop them or prevent them from making it public if they chose to do so.

A related risk is that circumstances are possible where another could create the same idea, and then patent it, preventing you from using your own idea. This possibility is somewhat decreased with the advent of the First inventor Defense amendment to the patent laws. However, the application of this defense is somewhat narrow. Only if you previously developed the same idea, did not derive it from the patenting party, and commercially used the idea more than one year before the effective filing date of the other party's patent will the defense apply. Importantly, it is limited to use of a method of doing or conducting business that otherwise would infringe a method claim. Thus, unless it falls within this narrow exemption, the risk still exists that another could patent the idea you have attempted to keep trade secret.

Of Broad Claims and Picket Fences

It is evident that the broader your claims, the more valuable the patent. How can one arrive there? Typically, the potential patenting of an invention comes up when an inventor has some interesting data that give a surprising advantage. Often one can quickly imagine claims that are directed to what the data show. It is very useful, however, to look at that particular example, and ask many questions on why it worked and to what other situations it can be applied. Ask: how did you come up with this idea? If there are ranges, what is the upper limit and the lower limit? How did you reach that conclusion? What else can this be applied to? By starting small, and working backwards, it is possible to see the whole of the invention from a small set of data. Ask about the known prior art. Know your borders and understand the theory of the idea and you have the potential to expand your claims.

For example, the scientists bring data that show it is possible to express the trypsinogen gene in plants. This is surprising because the plants didn't die; this protease would ordinarily eat up the plant while being produced. The first claim you discuss might be expressing trypsinogen in a plant. You conduct a search to find no one has done this before. But, why did it work? You discover upon questioning that it is because the pro-form of the protease was used rather than the active form. The protease has a chance to increase in the plant without killing it. Can this work with other proteases? It appears so; now you have a potential claim for expressing the pro-form of any protease in a plant. By exploring the theory and asking questions you have expanded the claims.

Similarly, if a protein is extracted preferentially at 50°C by incubation with a particular iron salt buffer, you can consider a claim with those perimeters. But why does it work better? The iron salt buffer appears to stabilize the protein so it can withstand a range of temperatures that it was sensitive to before. Will any metal salt work? What is the highest and lowest range now possible that was not achievable before?

Thus, it can pay to spend time writing claims over several times and asking many questions about how far this applies and where the point of invention lies.

Another theory in the approach to the building of an intellectual property portfolio is creating a "picket fence" in your niche area. With this idea, one obtains a number of patents, which may be of intermediate or narrow scope. While having less breadth, they still create obstacles that a competitor must avoid in order to work in the same area. There is much to be said for the picket fence approach, and indeed what appears to be a narrow claim may not practically be narrow at all. If, for example, you have a claim to a particular nucleotide sequence, but it is the only one that expresses in a plant with any efficiency to make it commercially worthwhile, then your claim is in fact quite broad in its impact. Keeping this approach in mind can give you layers of protection.

The varying types of protection you may want to consider include: the broad area of the science, as discussed above; your particular product, including the broad class of your products, the plant as a whole expressing the protein, the extracted protein (if new), the composition consisting of plant material and the protein; a new nucleotide sequence that encodes a protein of interest; the vectors used; components of the vectors used in producing the product, such as truncated versions of a nucleotide sequence, promoters, markers, leaders, or the like; processes for introducing the nucleotide sequence into plants; steps that improve expression, stability or purity in the plant; extraction steps; purification steps; and to the plant variety, just to name a few.

Building Research from an IP Point of View

Along these same lines, it is most beneficial to take time during a research project to discuss it from an intellectual property point of view and ask how it can be directed for maximum intellectual property return. There may be times when it is not practical to adapt research to consider patenting arguments, but at other times it may be relatively easy and make it much more likely that you can patent the invention.

You may meet with your scientists on the research described above, and pinpoint a theory as to why a process is working better or why you have been able to express particular nucleotide sequences in a plant that no one could before. By showing examples with other subspecies of the genus which defines your invention you can provide further convincing evidence to the Patent Office that the theory works and you are entitled to broader claims. As in the example above, if you understand that pro-forms of a protease can be expressed in a plant, try the experiment with two different pro-enzymes. If you believe other metal salts will work in improving recovery of the protein, try several other metals, and at differing temperatures. There is no particular number of examples that will necessarily convince the Patent Office that the idea applies across its genus. However, it can be generalized that one example is unlikely to be enough, and three may be convincing.

Further, if there is reason to question whether a process or introduction of a gene can apply across species, it may be valuable to demonstrate the invention in a monocotyledon and in a dicotcotyledon. For example, in *In re Goodman*, 11 F3d 1046 (Fed. Cir 1993) the Court found that although the claims originally filed were to mammalian peptides expressed in any plant, since there was only one example, to expression in tobacco and there was reason to question whether it would function the same in other plants, the claims would be limited to claims to dicot plants only.

In summary, it is worthwhile from an intellectual property perspective to step back from the research from time to time and take a look at it with the help of a patent expert. It can provide a perspective that allows for a few extra examples that may make patenting much easier.

International patents

Throughout the review above, international filings are mentioned, and it is wise to consider the impact of international law on your strategy. Whether you will have a keen interest in filing abroad will depend on several factors. For example, where will the product be produced or sold? Will any components come from another country or will any parts of the key processes take place abroad? Where will the product itself be imported?

For each such country where you have an interest, it is worthwhile to consider international patenting. This can add depth and value to your patent portfolio. One year from the U.S. filing of an application, one must decide where to file internationally in order to obtain the priority date of the national filing. Obviously, it is possible to file earlier, but the latest the decision can be made is one year after the U.S. application is filed. It is possible to file in each individual country where you have an interest, but the most efficient route is to use the Patent Cooperation Treaty filing. With this tool, one can obtain a priority date in over 100 countries by filing one application. Further, it has the advantage of allowing you to delay action in those countries if you are not certain how the invention will develop or if it is worth the money to file in the countries. Using this method it is possible to delay entry into the individual countries, and thereby delay expensive national costs, up to 20 months to 30 months from the priority date.

Once you reach the national stage, it can be quite expensive, considering translation costs, international attorney fees and related expenses. Before you make the decision to enter the national phase of a country it is worthwhile to first investigate the type of claims allowable. Some countries have considerable restrictions on subject matter that can be patented. Practical considerations include: what is the likelihood you could get favorable results if you had to enforce the patent; how much will it cost compared to the income from the product; could the product be imported into this country from a nearby country in which you do have an interest; and, would plant variety protection be cheaper and more practical. Also of importance is to consider working requirements in other countries. Many countries require the invention to be "worked", that is used within a certain number of years from grant of the patent (three to four years is common). If it is not worked, then it can be subject to compulsory licensing to others who apply to use the invention. This can make the expenses of the patent needless.

When filing a U.S. application, it is also important to keep international requirements in mind. The U.S. is the only country that is a "first to invent" country that rewards a patent to the first one to invent. Instead, nearly every other country is "first to file" meaning that it is the first person to the patent office who is given a patent. What this means is that in the U.S. if you reveal your invention through a publication, talk, or offer to sell to another, you have one year to file your patent application. If you do, you are eligible for a patent over anyone else who invented later than you. However, in other countries, if you publicly disclose the invention, you will not get a patent. The practical impact is that the one year grace period in the U.S. has no meaning if you wish to file abroad. Instead, scientists must not publicly disclose their invention before the patent application is filed. Further, if a biological deposit of DNA, callus, seed, or the like is required in order to adequately describe the invention, it must be made before the patent application under international rules. The U.S. allows a deposit of biological materials to occur after the application is filed, so that if it is necessary in order to "describe" the invention, it is available. Internationally, such deposits must be made first.

The Patenting Process: How it Works

The first step in patenting is to provide to the scientists a document they can fill out when they believe they may have a patentable invention. Training them on how to recognize patents will help them understand when they should consider the option. Also, having a person assigned to taking care of such issues (preferably a patent attorney) assures that surveillance for additions to the patent portfolio is underway. Typically, the scientists may be provided with a form that asks them to provide: a short description of the nature of the invention; what they know of prior work (prior art); how what they have done differs from what has come before; how they got the idea (this can be very instructive); who was involved and how (this helps with inventorship decisions); when the idea was developed; what further work needs to be conducted; any publications or offers for sale of the work.

Once this is completed, it is usually presented to a patent attorney who then will conduct a search for relevant prior art in the area. An opinion may then be provided on the chances of patenting the invention. Sometimes this step is skipped where the inventors feel they know the art well and that this is a new development. This risks that there could be unknown prior art, but saves costs on searching and opinions.

Next, a patent application is drafted and the attorney questions the inventors on the scope of the claims. The scientists must provide protocols and data to support the application. After the application is filed with the Patent Office, the waiting begins. Six months to a year later, the Patent Office will provide an Office Action. Nine times out of ten this is a rejection of the claims as not patentable for a variety of reasons. (Patent attorneys often wonder if they narrowed their claims too much if they receive the rare first office action allowance of claims). Then, the arguing with the Patent Office begins. The attorney writes a rebuttal, perhaps changing the claims and presenting additional data or other proofs. Eventually a "final" office action is received, which is final only in the sense that additional data may not be provided without permission. The case can be refiled at that time and the additional data presented. At some point, an interview with the examiner may occur, all with the view of narrowing the issues down to key points that the examiner considers important. If agreement can be reached, a patent will issue. Then, more applications may be filed from this patent in attempts to get additional and/or broader claims. Alternatively, if agreement is reached, an appeal can be filed to the Board of Patent Appeals, an administrative review to a three-examiner Board. In the past, such decisions have taken three years or more. If there is dissatisfaction at this point, further appeal is allowed to either the Federal Circuit Court, or District Court of Columbia, with final appeal to the Supreme Court, a rarity.

During this time, the patent has published if it is to be filed abroad. Under new U.S. and international rules, if someone infringes in the interim, when the patent finally issues, it is possible for the patent owner to go back and sue for past damages if the infringer infringed issued claims that fall within the scope of the claims as published.

Challenges to Patents and Enforcement

Once you have a patent, it is your right to decide who makes, uses or sells the invention as it is claimed. If another party refuses to honor that right you can sue for infringement. The tables can be turned on the patent holder, with the defendant asserting that the patent is not valid. In this instance the patent holder must bat 1000 each time they come up to the plate. They must prove the patent valid each time it is challenged. However, just one showing of invalidity is enough to defeat a patent. On the other hand, as noted, a patent is presumed valid. Tomes have been written on litigation including patent litigation. Briefly, litigation can be a long and costly process for both parties. With appeals, the process can last five years or more and cost more than a million dollars. With this in mind, some patent holders turn to arbitration. It is shorter, can cost less, but can also be quite uncertain in the results, with varying levels of chances to present evidence.

More often than not, the parties will work out an agreement themselves since there is much to lose. If you become aware that another is infringing, it can be a tricky situation. If you tell the alleged infringer they are infringing, and the threat of litigation is indicated, it is possible for them to bring a declaratory judgment to find your patent invalid, in their favored jurisdiction. Thus, it is important to bring the patent to their attention without threatening them, so that you can chose when and if to sue. Also, if you are aware of their activity, but choose to do nothing, if enough time passes, they can claim "laches", or that they reasonably relied on your lack of activity. Litigation counsel can advise on ways to avoid this; but, in brief it is generally best to act within six years of finding someone is infringing, or to send an appropriate carefully drafted letter on the subject. In addition, never make statements intimating they may be off the hook or they can claim estoppel should you decide later to sue.

What are other methods to challenge a patent? If you discover a patent or printed publication that shows the patent is invalid, and that information was not available during prosecution of the patent before the Patent Office, you can file for Reexamination. One handicap of this in the past was that the challenger had no further involvement in the process, and if the patent was reexamined favorably to the patent holder, the patent has an increased presumption of validity. Some involvement by the challenger is now allowed under rules being developed by the Patent Office. Also, a patent applicant can apply for reissue of their own patent if they find something is wrong with it, in order to bolster its validity.

The Importance of Lab Notebooks

It often seems like a small point to business owners, but lack of a good procedure on lab notebooks can come back to haunt a company. In one instance, millions of dollars worth of patents were lost because lab notebooks were kept in three ring notebooks, never witnessed and rarely signed.

Why might you need to document? First, you will want to record results for your own scientific purposes. Regulatory requirements may also relate to lab notebooks. For intellectual property, there are two major times when lab notebooks are important. One is in presenting data to the Patent Office to prove that your invention works or to overcome prior art. The second is to prove either to the Patent Office or the Board of Interferences, or to a Court, that you indeed were the first to invent. When another party claims they were the first to invent, and not you, they may challenge it through an administrative process in the patent office known as an interference, or through a court challenge. It may also come up if the Patent Examiner believes another was first to invent. In these instances, the lab notebooks can be presented as proof of date of invention. To "invent" one must have shown conception of the idea; in other words it must be concrete enough that all one has to do is carry it out. Second, it must be reduced to practice, which can be accomplished by experimentation, or by filing an application that adequately describes the invention.

In order to be useful for these purposes, a lab notebook must be "trustworthy" showing integrity and have independent corroboration, and be understandable to one skilled in the art. Integrity has meant that in the typical lab notebook, the pages are bound on the side, numbered throughout, so that if a page is ripped out, it can be seen, and no pages can be added. If any change is made to the text, it is lined through, not obliterated, so that all changes can be seen. If something is added, like a graph, it should be stapled or permanently taped in the book. The inventor signs and dates the lab notebook. Further, independent corroboration is provided by having the lab notebook signed and dated by a noninventor relatively soon after the entry. (This is preferably two weeks after the entry, but can be a month or more if the court and circumstances are forgiving). There are many variations on these procedures, and obviously, with the advent of computer lab notebooks, meeting these standards has become more uncertain. This review does not go into such detail here, but the company can help itself considerably by having a uniform lab notebook policy that is reviewed by a patent attorney on a regular basis to keep abreast of the changing dynamics. Court cases have reflected that even where a particular lab notebook does not meet the standards, where the company had a policy on lab notebooks, the evidence still might be accepted.

Freedom To Operate

Now we look at the other perspective of intellectual property: do you have title to the house and are the access roads clear?

Freedom to Operate is a thorny issue for any company: can I practice my processes and produce my product without running afoul of other's patents?

When To Be Alert

An important point is to alert your scientists when they need to know there may be freedom to operate issues. Among the times they need to know when to raise the flag is

- When they are embarking on an area of research that is new to the group
- When improving on another's work
- When using another's materials

- When they are aware it is a competitive area.

Sharing Materials

Any time biological materials come in to a company, it is always important to ask: is there a patent covering this material, and did the person who gave it to us have the right to pass along possession? This involves two aspects of freedom to operate issues: personal property and intellectual property.

In other words, the thing you were given is owned by someone such as the university, the company or the individual. Did the right person give you possession of the material? One way that companies attempt to solve this is to have the employer/university etc. sign off on all material transfers. Another is to have the person involved agree they have the right to give you the material. This will be useful only if they have actual or apparent authority to transfer rights to the material.

Second, someone may have patent rights to the material. It may be a different party than those who gave you the biological. This is worth checking out so you can proceed in comfort. Further, check the letter itself. Does the person transferring the material put any restrictions on your intellectual property rights to what you develop with the material?

Repeat FTO Studies

Outside of this situation, it is certainly possible to stumble onto another's patent area. Innocence, furthermore, is no defense. Thus, one is greatly incentived to check out freedom to operate at early stages of research. Understanding how wide and far to look for relevant patents can also be difficult. It is worthwhile to begin the process looking at the broad and working to the narrow. Further, it is important to repeat this process at several points.

For example, if you are embarking on new territory, it may be worthwhile to conduct what is sometimes called a "state of the art" search. These are very broad searches, which do not give great reliability in terms of being able to develop a specific product; however, they will alert you to major, important blocking patents. For example, perhaps you are interested in inserting nucleotide sequences encoding human growth hormone into plants. You will want to search to determine if the particular sequence you are using is covered by a patent and if there are broad patents to human growth human sequences. Has anyone ever inserted the DNA into plants? If you were involved in extraction and purification, you would want to look at those steps as well as the end product. This type of search will be far-reaching and not very specific. However, it would hopefully tell you whether you should avoid certain areas all together.

Examine Specifics and Broad Areas

As you develop a working process and product, again you will want to revisit the patent scene. The closer you are to a commercial product, the more intense and specific your searches and opinions should be. For example, it may include:

- Searches/opinions on the general area; for example, mammalian proteins expressed in plants, human growth hormones expressed in plants, industrial enzymes in plants
- The intermediate area: interferon expressed in plants, hepatitis protein expressed in plants
- Methodology of introducing the DNA into plants (e.g. *Agrobacterium* or bombardment)
- The general area of nucleotide sequences you are using (e.g. nucleotide sequences encoding human growth hormones of any type, etc.)
- The general area of proteins the sequences encode
- The specific nucleotide sequence you are using
- The specific amino acid sequence involved
- The promoters used
- The selectable markers used
- The leader sequences or other similar types of sequences used
- Terminator sequences used
- Extraction and purification procedures
- The end product itself
- Intended use of the product.

This is just an example to show the depth to which you should explore all aspects of your product and how it is made.

Options

No doubt you will unveil certain patent applications and patents that are an issue. When this occurs, you will want to further explore them.

For patent applications, they key question is: what is the scope of claims that are likely to issue? Many patent applications are filed and published with only marginal hope, at best, that the broadest claims presented will be allowed. In this instance, the patent attorney will review all public information about the application, and search for prior art. They will then attempt to answer the question of how the prior art constrains the claims and what claims might issue.

If the patent is issued then a thorough study of the patent will likely be conducted, including the file history. This is all the correspondence between the Patent Office and the prosecuting attorney. It is possible that during prosecution, statements were made that can impact the scope of the claims, beyond what is evident from a literal reading. Additionally, the attorney will look for other potential defects in the application. If the patent is invalid, an opinion can be rendered to that effect. The business person then has to make the decision as to whether to practice the invention, understanding the patent is invalid, and risk a lawsuit. The opinion will be valuable as a tool in avoiding treble damages. If it is well reasoned, it can be used to demonstrate lack of bad intent. Another option is to negotiate a license.

Still another option is to attempt to avoid the scope of the claims of the patent. The file history study will again be instructive here, since it is possible the claims extend only so far, either as read literally, or as reflected in the file history, and that with a few changes, one can avoid infringement.

SUMMARY

In summary, it is worthwhile to consider intellectual property from the beginning of the development of a company and to make it a priority in order to develop valuable assets and avoid trouble. Patents are the best protection available and are also hard to get and costly and time-consuming to obtain. Trade secrets are relatively inexpensive, last longer, but carry risks in terms of other's potential for use and the fact they are gone for good if they escape. Publication should also be considered as a defensive act.

If patent protection is sought, is wise to consider whether you have enough, or will have enough information and data within a year before filing an application. A provisional may be a good alternative if the data will be available within a year. When filing, do so before any public disclosure and if a deposit is needed, make it before filing if at all possible.

At the end of the year, consider filing a PCT application if international filing may be desired and delay the decision to file nationally where costs are a factor and there is uncertainty on how to use the invention. At the national stage, consider whether you will be using the invention in the country within three to four years and if it is worth the cost to go forward in a particular country.

It is wise to determine if you should prevent any publication of the invention before it is published for international purposes. Before publication at 18 months, again revisit the application to consider whether it should go on to publication. Carefully consider if claims need to be changed before publication so that provisional protection will capture what your competitors might attempt to practice before the patent issues.

Freedom to operate issues should be revisited at several points during research, and should be thoroughly reviewed when a final product is identified.

In general, intellectual property, when made a part of the overall process, can provide considerable value to a company. This can come from its own patent, trade secret and PVP portfolio, providing multiple layers of protection to the company's product and processes, as well as being able to present a picture of a clear path to product by resolving, as best as possible, freedom to operate issues.

A CASE STUDY

The following provides an example of how one might approach select freedom to operate issues in connection with development of a product involving protein production in plants. Keep in mind that at the time you read this, the technology and intellectual property issues that were relevant may not be now; it is an example of one approach.

In this example the research group at a company is interested in introducing into a plant nucleotide sequences that express the protein avidin. Avidin is a glycoprotein that has multiple uses. Avidin forms a particularly strong, non-covalent bond with biotin. It is this property of avidin that is primarily responsible for its commercial value, since it allows for detection of protein and nucleic acid molecules incorporating biotin. The customary source for commercial production of avidin has been chicken egg white, a method of relatively high production costs. Production in plants would be cost effective and eliminate certain concerns regarding contamination by animal pathogens and provides for improved stability of the protein during storage.

At this point prior work has described the avidin protein, and a nucleotide sequence obtained from chicken that encodes avidin. The scientists propose to start with this gene, optimize it for plant expression and introduce it into corn.

At this point a state of the art search is conducted and reveals only one relevant hit; a patent application that published, and had not been granted, to expressing avidin in plants for the purpose of providing insect resistance in plants. The application indicates that low levels of expression were obtained. Other journal articles on the subject show failure to obtain high levels of expression. There are no claims in the application to expressing avidin in plants per se. Thus, there appears at the time of the search to be no blocking patents in the subject matter area.

In addition to the broad state of the art search, a search is conducted for patents, applications or journal articles on the gene itself. The broad search is reviewed to determine whether there were any hits for avidin genes. The scientists are aware of two articles showing the cDNA sequence they intend to use as their starting material. The cDNA of the chicken avidin gene was documented by Gope *et al.*, Nucleic Acid Res. 15: 3595-06 (1987), and the genomic clone by Keinanen *et al.*, J. Steroid Biochem. 30: 17-21 (1988). In addition, Keinanen *et al.*, European J. Biochem. 220: 615-21 (1994), had identified a family of closely related avidin genes. The searching expert includes the names of all the authors of these articles in their search, as well as the institutions for which they work. No patents or applications are revealed in the search.

The freedom to operate analysis further includes assessment of the ability to use the avidin gene in the scientists' possession. The licensing department confirms that a material release agreement which accompanied transfer of the gene to the scientists did not impose any third party ownership or licensing issues on work with the gene, and was signed by a person within the institution having a right to transfer the material. It is determined that the other materials and processes to be used in the research have been previously studied for freedom to operate issues.

Later, the scientists return with news that they have been able to obtain high expression of the gene, attempts which have not been successful in the past. A patentability analysis is conducted, and it is determined no new patents or applications have appeared in the interim, and that there is a reasonable chance of obtaining patent protection. Since stable transformation has been proven, a patent application is filed.

Meanwhile, a third party has become interested in a potential commercial relationship for plant production of high-expressing avidin. Those in charge of product development give thought to making concrete what processes and components will be used given the currently available technology, what the end product will look like and how the product will be used. (Freedom to operate issues regarding use may be best addressed by the third party user). All this is reviewed with the patent attorney to determine if there are any freedom to operate issues. The following table was found to be a helpful starting point for discussions, which included review by the scientists, business development personnel, third party collaborator, all with review by the patent attorney.

The following is representative of the types of issues one may look at in reviewing possible freedom to operate issues relative to production of a "molecular farming" product. It may not be complete for your purposes and technology may have impacted what should be considered – it is an example only.

Product Review – an example

Product / Process	Specific Component / Step	Intellectual Property	IP / license status	Personal property issues?
Selectable marker				
Promoter				
Gene				
general				
sequences				
use				
Terminator				
Gene of interest				
Promoter				
Gene				
general				
sequences				
Signal sequence				
Leader sequence				
Terminator				
Selectable marker (optional)				
Transformation method				
General method				
Species specific				
Culture methods				
Regeneration				
Targeting				
Other steps/processes				
Vectors/vector development steps				
Vector				
Process developing vectors				
PCR or related				
Taq polymerase				
Others?				
Host plant				
Hybridization methods				
End product				
Use of end product				

SUMMARY

In summary, it is worthwhile to consider intellectual property from the beginning of the development of a company and to make it a priority in order to develop valuable assets and avoid trouble. Patents are the best protection available and are also hard to get and costly and time consuming to obtain. Trade secrets are relatively inexpensive, last longer, but carry risks in terms of other's potential for use and the fact they are gone for good if they escape. Publication should also be considered as a defensive act.

If patent protection is sought, is wise to consider whether you have enough, or will have enough information and data within a year before filing an application. A provisional may be a good alternative if the data will be available within a year. When filing, do so before any public disclosure and if a deposit is needed, make it before filing if at all possible.

At the end of the year, consider filing a PCT application if international filing may be desired and delay the decision to file national where costs are a factor and there is uncertainty on how to use the invention. At the national stage, consider whether you will be using the invention in the country within three to four years and if it is worth the cost to go forward in a particular country.

It is wise to determine if you should prevent any publication of the invention before it is published before international purposes. Before publication at 18 months again revisit the application to consider whether it should go on to publication. Carefully consider if claims need to be changed before publication so that provisional protection will capture what your competitors might attempt to practice before the patent issues.

Freedom to operate issues should be revisited at several points during research, and should be thoroughly reviewed when a final product is identified.

In general, intellectual property, when made a part of the overall process, can provide considerable value to a company. This can come from its own patent, trade secret and PVP portfolio, providing multiple layers of protection to the company's product and processes, as well as being able to present a picture of a clear path to product by resolving, as best as possible, freedom to operate issues.

INDEX

E.E. Hood and J.A. Howard (eds.), Plants as Factories for Protein Production, 207–209.